Henry Coppée

Elements of logic

Designed as a manual of instruction. Revised Edition

Henry Coppée

Elements of logic

Designed as a manual of instruction. Revised Edition

ISBN/EAN: 9783337277994

Printed in Europe, USA, Canada, Australia, Japan

Cover: Foto ©berggeist007 / pixelio.de

More available books at **www.hansebooks.com**

ELEMENTS OF LOGIC;

DESIGNED AS A

MANUAL OF INSTRUCTION.

BY

HENRY COPPÉE, LL.D.,
PRESIDENT OF THE LEHIGH UNIVERSITY.

REVISED EDITION.

PHILADELPHIA.
T. H. BUTLER & COMPANY.
1882.

Entered according to Act of Congress, in the year 1857, by

E. H. BUTLER & CO.,

In the Clerk's Office of the District Court of the United States in and for the Eastern District of Pennsylvania.

Entered according to Act of Congress, in the year 1872, by

E. H. BUTLER & CO.,

In the Office of the Librarian of Congress, at Washington.

WESTCOTT & THOMSON,
Stereotypers, Philada.

SHERMAN & CO.,
Printers, Philada.

PREFACE
TO THE REVISED EDITION.

IN obedience to the public demand, the publishers have spared no expense in giving to this volume a new and more attractive form. The author has, on his part, revised it carefully, and added much important matter, some of it embodying the valuable suggestions of instructors who have been using it for many years. Parts of the subject have received fuller illustration. Parts have, after careful deliberation, been omitted, and a chapter has been added on the Fundamental Laws of Thought or First Principles of Reason. The plan and divisions of the work remain the same. As many and apparently conflicting views have been taken of the meaning, genus and scope of Logic, as a branch of Philosophy, it seems proper to say that much of the diversity is nominal; that, with differences in name, most treatises admit the same functions of words, conceptions, propositions and arguments, and that the chief antagonism arises from an undue exaggeration of the place and value of certain functions in the reasoning process. This remark is made in the interest of those who are deterred by the apparent antagonism of systems from the study of a science most of the details of which are the same in all systems; the *body* of logical doctrines recognized by all logicians do not refuse to combine harmoniously in one system.

In the present edition the author has availed himself of the voluminous and exhaustive treatise of Sir William Hamilton, in which, together with the expression of his peculiar views and criticisms, some of which may be dissented from, the functions and the history of Logic have been set forth with great acuteness and erudition. This has led the author to slight modifications of the system of Whately, but none that will affect the general importance and soundness of his views. The numerous more recent treatises on the science have also been examined.

It is confidently hoped by the author and the publishers that the favor so continuously displayed towards this work ever since its appearance fifteen years ago will be increased by its additional value and its clear and attractive form; and that a subject frequently regarded as both abstruse and vague will be commended, by the clearness and simplicity of its treatment, to many who have been heretofore doubtful of its utility.

H. C.

THE LEHIGH UNIVERSITY, *August* 1, 1872.

PREFACE
TO THE FIRST EDITION.

THE following treatise has been written in the hope that it may supply, in some degree, a real want. For several years the author was a teacher of Logic in the Military Academy at West Point, where the subject was thoroughly studied by the aid of Archbishop Whately's text-book.

How much a manual was needed before that work appeared may be known from the significant fact that, as soon as it was published as an article in the Encyclopædia Metropolitana, it was eagerly caught at by the community of teachers, and used, unaltered, as a book for college instruction, on both sides of the Atlantic.

Since the publication of that article many have attempted the preparation of a manual which should have the instruction of classes as its original design; but the soundness of Whately's views and the conciseness of his expression still gave to his work the greatest circulation. Among so many endeavors the author would venture to express the hope that his little manual may find its special purpose and mission. It is short; it is explanatory of all the difficult points so often left to confuse a student; the arrangement is simple, and much that in a larger treatise would be of necessity included is here omitted, so that what the student learns in the limited time of a college term he may learn well, and retain in his memory as a basis for further investigations. To some persons it may seem too much simplified; but let it be remembered that it is a manual for youth, and that its only aim is to teach them the *Elements* of Logic as the foundation of all reasoning.

The basis of the work is "*Whately's Logic*"; many of the examples are taken directly from that; so many, indeed, that the acknowledgment is here made for them all, and for much that is excellent in arrangement and in expression. As the clear expounder of Aristotle, and the originator of much that is valuable, Whately must stand at the head of the Logicians of this age. The author would refer specially also to the material assistance obtained from "*Devey's Logic*" (Bohn's series); "*Aristotle's Post and Prior Analytics*" (Bohn's translation); "*Neil's Art of Reasoning;*" "*Blakely's Historical Sketch of Logic;*" "*Lord Bacon's New Organon;* Arnauld (Logique de Port Royal); *J. Bentham's* "*Book of Fallacies.*" From *Neil* a few of the examples have been taken.

Besides these, he has consulted a great number of works, the aid derived from which is so general that they do not require special mention.

UNIVERSITY OF PENNSYLVANIA, *July*, 1857.

TABLE OF CONTENTS.

CHAPTER I.

LOGIC, THE MEANING OF THE TERM, AND THE SCOPE OF THE SCIENCE.

Section 1. Of the term Logic.. 13
   2. Sources of Error.. 14
   3. Logic and Philosophy.. 15
   4. Logic and Rhetoric... 17
   5. Objection to Logic as an Art................................. 18
   6. Natural Logic.. 20
   7. Systematic forms of Error..................................... 21
   8. Of Method.. 22
   9. Analysis and Synthesis... 23
   10. Analysis and Synthesis as applied to Logic......... 27
      1. Analytical View. 2. Synthesis....................... 27
      3. Historical View... 28

CHAPTER II.

ANALYTICAL VIEW OF LOGIC.

Section 11. The Reasoning process Analyzed................... 29
      The Dictum of Aristotle.................................... 31

CHAPTER III.

A SYNTHESIS OF LOGIC.

Section 12. Of certain operations and states of the Mind used in the process of Argument... 35

1. Apprehension.......... 35
2. Judgment........ 36
3. Reasoning............. 37

CHAPTER IV.

Section 13. Of Terms.......... 39
14. Division of Simple Terms....... 40
15. Quantity and Quality of Terms........... 42

CHAPTER V.

Of those Operations which Relate to Terms.

Section 16. Abstraction and Generalization.......... 44
17. Species, Genus and Differentia............ 45
18. Property and Accident........... 47
19. Of the different orders of Genera and Species............ 48
20. Realism and Nominalism........... 50
21. Definition of Terms........... 51
22. Nominal and Real Definitions............ 53
23. Rules for Definition............ 54
24. Division......... 58
25. Recapitulation......... 62

CHAPTER VI.

Section 26. Propositions............ 63
27. Simple and Compound............ 66
28. Quantity and Quality of Propositions......... 67
29. Of the Distribution of Terms in Propositions............ 70
30. Conversion 73
    Illative Conversion............ 76
31. Of Opposition 78
32. Of the Matter of Propositions............ 79
33. Of Compound Propositions............ 81
34. The New Analytic............ 84

TABLE OF CONTENTS.

CHAPTER VII.

Section 35. Of Arguments.. 87
36. Of the Syllogism... 88
37. Logical Axioms... 89

CHAPTER VIII.

Of Figure and Moods.

Section 38. Figure... 95
39. Mood... 98
40. Of Reduction.. 107
41. Indirect Reduction... 111
42. Notation of the Syllogism.. 113

CHAPTER IX.

Of Irregular, Informal and Compound Arguments.

Section 43. Of Abridged Syllogisms....................................... 118
44. The Sorites... 121
45. The Epichirema... 124
46. Of Hypothetical Syllogisms...................................... 126
    Conditional 126
    Disjunctive 130
    The Dilemma... 132

CHAPTER X.

Fallacies.

Section 47. The Meaning and Comprehension of a Fallacy......... 137
48. Fallacies *in dictione*.. 138
49. Material or Informal Fallacies.................................. 141
    Errors in the Premisses... 142
    Petitio Principii 142
    Arguing in a Circle... 142
    Non causa pro causa.. 143
    Errors in the conclusion....................................... 145

	PAGE
Irrelevant conclusion	145
Argumentum ad rem, etc.	148
Changing the point in dispute	148
Fallacy of Objections	149
50. Verbal Fallacies	150
1. Etymology	152
2. Interrogations	153
3. Amphibolous Sentences	154
Causes of Ambiguity	155
51. The manner of removing Ambiguity	160
52. The Fallacy of Probabilities	161
53. Popular Fallacies	163

CHAPTER XI.

THE FUNDAMENTAL LAWS OF THOUGHT OR FIRST PRINCIPLES OF REASON.

1. Identity	167
2. Contradiction	168
3. Excluded Middle	168
4. Reason and Consequent	168

CHAPTER XII.

Section 54. Of certain modes in which Logic is applied	173
Intuition, Induction, Deduction	173
Argument *a priori*, *a posteriori*, *a fortiori*	176–178
The Investigation and Discovery of Truth	178
1. Observation	179
2. Hypothesis	179
3. Induction	179
4. Theory	180
Of the Nature and Kinds of Evidence	180
Consciousness	180
Sensation	180
Analogy	181
Induction	181
Testimony	181

CHAPTER XIII.

A Historical Sketch of Logic.

Section 55. Division of the Subject.................................. 182
    1. Aristotle. 2. Christianity and Logic. 3. Bacon and the rise of Inductive Science. 4. The Present System.
56. Aristotle... 184
    The Categories.. 188
57. The Logic of Christianity.................................... 193
58. The Logic of Experimental Philosophy................ 200
59. Logic in the 18th and 19th centuries.................. 209
60. Categories and Classification............................. 211
61. Conclusion... 217

APPENDIX.

Examples for Praxis... 218

LOGIC.

CHAPTER I.

LOGIC: THE MEANING OF THE TERM AND THE SCOPE OF THE SCIENCE.

(1.) Of the Term Logic.

LOGIC is directly from the Greek λογική, feminine of the adjective λογικος, and implies ἐπιστήμη or τέχνη—*science* or *art*. The adjective is from the noun *logos*. As, of all the Greek words which have been transferred to our English speech, none is vaguer and more subtle in its meaning than the word *logos* (λογος), so, of all the sciences, none has been less clearly defined, both as to its meaning and its scope, than the science of *Logic*, the name of which is taken from that word; and, in consequence, no term is more erroneously applied and more frequently misapplied than the name itself.

Logos means both *thought* and *speech*, and the earlier writers distinguish it as being both *that in the mind* and *that without*. Combining these, *logos* came to mean *discourse*, and hence some writers have supposed *Logic* to be simply the science of spoken or written *language*, thus confounding it, in part, with Rhetoric, and even with Grammar; others, considering *discourse* to imply not simply the *written symbol* or the *spoken sound*, but also the *expression of the thought*, have more correctly supposed Logic to be the *Science of the Laws of Thought*, and, as such, a branch of metaphysics, or the science which investigates the workings of the mind; others still, and by far the greater number, regarding it as a union of language and thought in the deduction of truth, have claimed that it had to do with the *subject-matter of scientific investigation*, and have thus erred more widely than all by confounding Logic with the labors of physical, metaphysical and ethical philosophy rather than an instrument for the service of them all, as it really is.

It seems necessary, then, at the beginning of a treatise on this subject, to define the meaning of the word, and the true scope of the science, before we undertake its study—to rid ourselves, as it were, of the mists which surround us, before we can even see clearly the field in which we are to labor.

(2.) Sources of Error.

Many accurate thinkers have confused the minds of students by producing books which, while they contain a just view of the *logical system* itself, attempt at every step, as has been said, to explain the *subject-matter* upon which this system is employed, and which forms no part of it; while many others, adopting strongly the views of those who have initiated so-called *systems* of logic, have, as partisans, carried forward from period to period old errors and old perplexities; and, themselves ignorant of the subtleties which surround them, have called *their* views the *true logic*, and those of every other writer *false*. Others again have endeavored, in an amiable but unscientific spirit, to harmonize all the schemes of the philosophers, and to call the result, full of error and inexactness, *the system of Logic*.

There are, indeed, in the systems of the great philosophers many parts that are mutually dependent, and true science will be found to harmonize with itself everywhere. But since there is also error in them all, no mere greatness of name should exempt from the scrutiny and exposure of error.

We must take care to distinguish between the different functions of the intellect, so as to call things by their right names—not including in the name Logic what belongs to Physics or Metaphysics, but laying down at the outset the limits and province of that system which we wish to designate by the word Logic. If we can do this we shall have accomplished very much at the beginning, and shall find our labor easy as we proceed.

If we would see how important it is rightly to understand this fact of the ambiguity of the word Logic, as frequently employed, we need but look for a moment at the errors into which modern philosophers have fallen when speaking of the *Logic* of Aristotle as compared with the *Logic* of Bacon. This has been fostered by the fact that while Aristotle set forth his logical views in his *Organon*, Bacon produced a *Novum organum* or new organon. If, as we shall endeavor to demonstrate, Logic is the science which controls the universal and ultimate principle of reasoning, given to man, just as speech was given to him, by a beneficent Creator, then it is not *Aristotle's Logic*, nor *Bacon's Logic*, but a single universal Logic, given to man as the rule of his reason, which must be intelligible and harmonious wherever and by whomever it is used.

(3.) Logic and Philosophy.

In this consideration another word plays a prominent part. The word which has been pressed into service, to denote the peculiar progress of great minds in the domains of Truth, is "*Philosophy;*" but even the word "Philosopher," said to be adopted by a wise ancient* as a more modest title than σοφος, as the sages of Greece were called, has been productive of great confusion. "Philosophy" has been made to stand for a thousand sciences, and to preside in the kingdoms of mind, morals, and physics, until to be a *philosopher* means to pursue one of many intellectual pursuits, and *Philosophy* unqualified means everything or nothing.

* Pythagoras.

And yet this vague and inexact term *Philosophy* is the one which has been most frequently confounded with *Logic*, and a want of clear definition and of a just understanding in the dispute has led to the production of inexact, distorted, and conflicting *systems*, both of *Philosophy* and *Logic*, which have confused those desirous of learning, and deterred many from the difficult and perilous attempt. In attempting to reach a clear division and definition of Philosophy and Logic, the followers of Plato asserted Logic to be a part—and the instrument—of Philosophy. The Stoics divided Philosophy into three parts, viz.: *Physics* or Theoretical philosophy; *Ethics* or Practical philosophy, and *Logic*, a subsidiary part, instrumental to the others.

Indeed both words, and the errors to which their use has led, indicate, at once, *the yearning* and *the weakness* of the human mind—the desire of man to investigate and systematize truth, combined with the obscurity and doubt which beset his investigations at every step.

The acuteness of the Greeks, upon which had been grafted all the power and attainment of the Oriental world, could reach no clearer nomenclature than to call their studies and their inductions *Philosophy*—the *love* rather than the attainment of *wisdom*—and the art by which they reasoned from truth to truth, by which they progressed from parallel to parallel in the sea of doubt and uncertainty, *Logic*, the art of *words* or *discourse*, the very mention of which suggests a dubious question, and calls up, as it were, two opponents in considering it.

Without considering the numerous definitions, we may agree to call *Philosophy* a search for final causes, in accordance with a primary law of the mind, which demands a cause for everything, and also in obedience to the tendency of all science to unity. This covers the investigation of truth as to its subject matter; the processes of collating and comparing material, and of classifying and aggregating observations and experiments.

Logic we shall consider the science which guides the operation of thought from simple intuitions and conceptions, through judgments, to the simple reasoning process, by which we pass from truth to truth already found, and by which we guard against fallacious arguments in the passage.

(4.) Logic and Rhetoric.

The exact line between Logic and Rhetoric is not always clearly drawn. The distinction between them may be thus stated: *Rhetoric* is the art of inventing, arranging and expressing thought in discourse, or, in brief, it is *the Art of Discourse*. Rhetoric *finds* terms, propositions and arguments in the construction of discourse, and arranges and clothes them with language to produce a certain effect.

It is the province of *Logic* to test the Rhetorical operations, and particularly to declare of its arguments whether they are *valid or invalid*. Thus, in its relation to Rhetoric, Logic is a check and an ordeal; an arbiter of the reason; a detecter of what is false and fallacious.

In this view Rhetoric includes Grammar. Thus a discourse may be grammatically correct, and rhetorically elegant, and yet full of error as to its Logic.

Having thus seen that the name Logic is in a great degree arbitrary, and that we should not attain to an understanding of the subject, if we followed, even remotely, the etymology of the word, we repeat that Logic has to do neither with the words themselves—except as they are arranged into terms, *propositions* and *arguments*—nor with their meanings, except as related to *reasoning*, i. e., *passing from two known and acknowledged judgments to a third, which is derived from their combination*. With this explanation, then, we may state the definition of the term. Logic is *the Science and the Art of Reasoning;* and reasoning is the ultimate process of thought in its search for the *True*, the end proposed to us by our cognitive faculties.

Of these two terms, *Science* and *Art*, we remark that *Art* is in a critical sense more extensive than *Science*, since the *practice of an Art* implies the application of the *principles of Science*, while, on the other hand, Science might, indeed does, exist in its *theoretic* state without being put to practical use. The Science would be the investigation of the principles upon which the human mind is based in reasoning, and the *Art* the application of those principles to the establishment of practical rules for conducting the process. Logic may then be more simply defined the *Art of Reasoning*, and as such we shall consider it in these pages, less concerned about the composition of man's reason than about the practical laws and methods by which it works.

Before proceeding to explain the system of Logic, which has developed itself since the days of Aristotle, let us meet at the threshold some plausible objections which have been brought against the establishment of any system whatever.

(5.) Objection to Logic as an Art.

As man has been universally gifted with reason, by means of which he may combine his thoughts and arrive at just conclusions, and with language in which to communicate them, it is asserted that every man carries his own Logic within him, as the immediate gift of God.

All men reason, it is true, and many men are not aware of the logical process which they use; and this has been made, even by men of acute minds, an objection against Logic; for, they say, since men reason, and reason well, without rules, and without knowing the process, a system of rules must be unnecessary.

The objection is plausible, and has been fruitful of evil. But as it is one which may be brought against many other arts as well as Logic, it may, we think, be most easily met and most clearly refuted by illustration. Many children speak with correctness and precision before they have any knowledge of Grammar; and there are persons of wonderful powers in arithmetical computation who have never learned Arithmetic; but *Grammar* and *Arithmetic* are not for such reasons condemned: their rules are an infallible test for *precise speaking* and *correct computation*, and are thus guides to the weaker and slower intellects—and these constitute the immense majority of mankind—to keep them from formal error. So, too, in Music and Painting; great geniuses arise in both Arts, but no one would contend that hard study, according to the established systems of the great composers and the great masters—established upon the true principle of voice and ear and eye—is not absolutely requisite to excellence and success.

Many persons of clear perceptive faculties, and who form and combine their judgments rapidly, may reason acutely and well without a system of rules; but, in order to be certain of their correctness, others must have some invariable test; on the other hand there are many, of quick but erratic minds, who reason with such dangerous sophistry that the most delicate logical tests alone can expose the fallacy, of which indeed they may not themselves be entirely aware. As such delicate tests have not been within the reach of the multitude, it is thus that men have become, for want of a popular knowledge of Logic, at once self-deceivers and deluders of mankind: have established illogical religious creeds, monstrous social fallacies, false theories of government, which are immediately made manifest by the simple application of Logic.

Nay, more: since Logic is the science which develops the one universal principle of Reasoning, applied alike to every branch of science, Exact or Inductive, it seems much more necessary that we should establish full and unerring rules for our guidance, and thus be kept, at every turn, from the manifold errors which arise from systems based upon such objections as those we have mentioned.

(6.) Natural Logic.

The natural laws which govern the human mind in its attempts to reason have been called by the opposers of Logical systems *Natural* Logic. We accept the name, and are ready to allow that, in following these laws, reason is right, and originally perfect in applying them; but now, in the fallen condition of man, reason is certainly liable to be biased by prejudice, distorted by passion, or insidiously tempted into open error. Thus many men, who reason correctly on most subjects, are swayed, in one or more, by self-interest, partisanship, fashion, predominance of the imagination, and such like causes; and thus men of equally clear minds in the main, from the same premises draw different conclusions, or establish the same conclusion by very different premises. Thus also the same man, at different periods of his life, or swayed by various circumstances, will reason differently; and from such causes, it is evident that each man's natural Logic is not a sufficient guide for his reason. Besides, reason does not confine itself to the immediate conclusion flowing from these fundamental laws of reasoning, but is constantly drawing one conclusion from another. Now, in this process, reason certainly needs more than these natural laws to keep it from error.

Yet still it is from this natural Logic, or, rather, the concurrence of the right reason of many well-ordered minds, that the science of Logic has been deduced.

By a systematic observation of such minds, as they reason, taking care to remove all causes of error in each particular case, we establish rules for the reason, and are able to detect, by the application of these rules to other cases, every fallacious argument resulting from such causes of error.

There must have been reason before there could be a system of laws to govern it, just as we know there was language before Grammar was formed. It was to systematize this reason, to methodize this natural Logic, and particularly to guard against errors in the use of the reasoning powers, that a canon was prepared, and that a complete science of Logic has been formed.

We have spoken in general terms of the confusion and error which have grown out of the misapprehension of Logic. The more special phases of it are those resulting from an attempt to systematize these general erroneous notions.

(7.) Systematic Forms of Error.

By a very common misuse of language, we hear such phrases as "*mathematical reasoning,*" "*moral reasoning,*" "*syllogistic reasoning,*" and "*inductive reasoning;*" which would lead us to suppose that instead of *one* there were *many* kinds of reasoning. This is a fruitful source of error.

These so-called different kinds of reasoning are only applications of Logic to different subjects and different habits of thought. The *Logic* in each is the same; the subject-matter alone is different.

It would seem unnecessary to dwell upon this point, but it has been so commonly misunderstood, and the error has been so disseminated by professional writers upon Logic, that it must be plainly stated and carefully remembered.

When we speak, then, of a good mathematician, we mean one who is able, most surely and rapidly, *to apply Logic to the investigations of numbers and quantity.* When we hear of a great theologian, we know that he has amassed much theological learning, and *has applied Logic to it successfully.* So, too, with other sciences.

In general, in whichever of the myriad fields of nature and mind ardent votaries may wander, however various the stores they may amass, they must all come back with their sheaves to the great measuring-centre of Logic, and apply its dicta before they can compute or use their gathered gains.

The value of Logic as a study is manifold. Not only is it an infallible test of argument, but it strengthens and disciplines the mind, giving it system and method; and it has established a terminology of universal adoption and applicable to all its practical adaptations in science. Thus it gives uniformity to the investigation of all branches of science.

(8.) Of Method.

Method is the order and arrangement of facts to produce a certain result; to establish new truth, to investigate old, and to explain and teach both. It is derived from the Greek μεθ'οδου, which denotes the *way through which* we arrive at a certain result. Method is employed in every science, and plays a specially important part in Logic.

Whatever steps are taken to make knowledge profitable, to reduce theory to practice, and to give clear, distinct and connected ideas of science, constitute Method. The extension of the term *Method*, it is evident, will differ according to the subject to which it is applied.

The *methods of investigation* differ slightly for the different kinds of science, but may generally be classified under two heads, *Analysis* and *Synthesis*, of which the former is generally used in the private investigation of truth, and the latter for the purposes of instruction.

The successive stages in the discovery, progress and establishment of any science are three, viz.: the *descriptive*, the *inductive* (also called the experimental), and the *deductive* or exact stage.

As soon as, by the *description* of a science, the statement of its present condition, its wants, its unknown causes, etc., we have a just representation of it, we proceed to observation and experiment, or *induction;* and when, by *induction*, or the labored collection of many particular facts and examples, we have established *general laws*, we may then *deduce* from them any particular fact or facts which it concerns us to know.

These stages of investigation belong equally to the *physical* and *moral* sciences, with the slight difference in practice that the vagueness and complexity involved in mental, spiritual and social phenomena, which all belong to the moral sciences, require more delicate and subtle agencies to trace their laws than those of the natural world around us.

And the sources of experiment are not at all analogous. Here we are surrounded by apparent contradictions. The world of nature is changeable and shifting, and yet it is palpable to our senses; the laws which govern it are mysterious and inscrutable, and yet they are constant; the moral world, which is unchangeable and eternal, is, when considered or examined by unaided reason, vague and obscure, and the abstract conclusions to which our inductions lead us, positive and incontrovertible as they are, are but few and unsatisfactory.

We shall have occasion to consider the subject of Method more in detail hereafter, but at present we design to apply it to the consideration of Logic.

We speak of the method of a single science, or a Method which is applied to all—as in that which leads to the Classification of the sciences. In either investigation the division of Method into Analysis and Synthesis is a just one, as both are used in either process.

(9.) Analysis and Synthesis.

To illustrate more clearly the nature of these two processes, let us take a familiar example. If we designed to teach a person how to make and use some complicated structure, as, for example, a ship, and if this person had never seen one, the first step in the process would be to show him the ship completely built and ready to proceed to sea, fully rigged, equipped and manned, that he might take in at a glance its finished appearance, and its ultimate design and use: in a word, that he might know *what* he was to learn to make. This would be the first lesson in ship-building. The next step would be to show it to him partially dismantled, or, in effect, to take it to pieces before his eyes, that he might see the parts of which it is composed, and their relative position in the structure.

The third step would be to show him how each part was made, and to let him see them all in minute detail lying together, according to some system, which should be preparatory to a reconstruction of the ship.

This process of successive steps is *Analysis*,* or a dissolution of anything into its elements.

In the investigation of any science, it is of primary importance. Showing us at first the scope and design of the science, by systematic degrees it decomposes it into its elements, and prepares us for intelligent study of its many forms.

This operation shows us also the simplicity of science, and is evidently derived from the teachings of nature; for, while there are innumerable forms of animal and vegetable life, the analysis of nature which is constantly going on shows but few parts or elements in all her works, and great simplicity of combination of the same elements in different proportions, to produce the most dissimilar forms and results. So all the sciences, physical, intellectual, and moral, while they assume many and varying forms, are in reality composed of a few simple elements of nature or mind, and this their analysis displays.

The analysis of physical science is of course the most exact of these processes, in proportion as the things of sense are easier to comprehend and fix than those of mind and spirit; in physics, this process of analysis is carried from the grandest class, such as kingdoms and high genera, to the observation and use of atoms and molecules inconceivably small, which

* αναλυω—to separate into elements.

are to constitute the basis-elements of a reconstructing process. Accurate analysis is a work of patient labor. Chance experiments have indeed occasionally produced great results, but this is an argument for, rather than against, careful analysis. Roger Bacon discovered a fulminating powder when he was not seeking it; but, to be useful, this powder must cease to be a chance discovery; that is, it must be analyzed into *nitre, charcoal,* and *brimstone,* so that, these constituents once known, we can make our fulminating powder at will. Science has never proceeded upon chance; it moves safely only when it moves by invariable but ever-extending laws.

Incomplete analysis has done more to establish and perpetuate error than even blind superstition. For it was in the face of the latter that Copernicus and Galileo established the true theory of the heliocentric system; while, before their time, the incomplete, false, and arbitrary analysis of astronomy, and the belief in stellar influences, which a just analysis would have destroyed, led all the writers, from the time of Ptolemy, to build a false system of celestial mechanics, and thus to clog the wheels of true science.

The process of analysis having been completed, we come naturally to *Synthesis.**

Having taken to pieces, we proceed to the other task of rebuilding: carefully examining each different element as they all lie before us, until we understand thoroughly the material of which it is made and its construction, we proceed to adjust it to its place in the structure; piece by piece, perhaps slowly and painfully, we build the ship, until at length it is complete; nor is the labor yet finished: we launch it upon the waters, spread its sails to the wind, and see it in practical and successful movement, and then we may account ourselves acquainted with the structure, and able to build its like whenever called upon to do so.

* συντίθημι—to place together,

This operation is called Synthesis; it is evident that it is also continually going on in nature in the reproduction out of crude materials of the many forms of complicated existence.

Many writers, in investigating a science, begin with this latter process, entirely neglecting the former; but it is so evident that the analysis of a science gives large and valuable lessons preparatory to its synthesis, or real study for ourselves, that most modern treatises on science have adopted and followed this order of instruction. It may then be safely stated that in any science the true synthesis can only be proportional to a vigorous and just analysis, and there have consequently been rules laid down for proceeding to consider any science or art in pursuance of this method.

The rules for Analysis may be reduced to these:

1st. Not to believe any general scientific statement without proof; that proof determined by the just principles of evidence.

2d. To divide every scientific dictum into as many parts or elements as shall be necessary to resolve it.

3d. To make a methodical arrangement of these elements in order that we may understand them clearly and the relation which they bear to each other.

Having done this, the corresponding rules for Synthesis are:

1st. To use such *terms* to express the elementary parts as are free from ambiguity.

2d. In combining these, to assume only such clear principles or *axioms* as cannot be contested by any persons.

3d. To prove, by demonstration, all the conclusions at which we arrive, in the employment of the terms and axioms used.

These remarks upon analysis and synthesis, as the two vital functions of Method in investigation, and as the two necessary instruments of all scientific study, are designed for general application. A proper and constant application of the rules

ANALYSIS AND SYNTHESIS AS APPLIED TO LOGIC. 27

of analysis and synthesis would cause great advancement in our studies, and would go far to insure us from error, however rapid that advancement might be. Analysis and synthesis are conducted by means of abstraction, generalization, definition and division, which will be referred to hereafter. We have placed the subject of Method in this place, because we design to use it in application to the study of Logic itself; for, as a science to be studied, Logic comes under the rules which have been just laid down.

(10.) **Analysis and Synthesis as applied to Logic.**

Now, let us employ this method in investigating the science of Logic.

Abstract or *formal logic* is an explication of the laws of thought and the rules of reasoning, without regard to any subject-matter. *Applied logic* is the application of these rules to the subject-matter of scientific investigation. It is only with the first of these that we at present have to do.

That we may study the subject profitably, making each step a preliminary to the due understanding of the successive steps, we propose to divide the entire subject into the following special considerations:

1. AN ANALYTICAL VIEW OF LOGIC.

In this we regard the science in its aim and its workings, and after thus showing its design and its scope, we analyze or dissolve it into its different parts, showing what those parts are which effect by their combination the purpose designed.

2. A SYNTHESIS OF FORMAL LOGIC.

As Synthesis is the reverse process of Analysis, and as an Analysis of such a study would be in reality but a general view of the scope of that science which Synthesis is to establish, we shall see that while our analytical view of Logic may be brief and general, our synthesis must be minute and careful. We must more particularly examine those parts which our analysis has given us, in order that we may be able duly to combine them in their just relations.

In imparting instruction upon subjects which are known, the synthesis is evidently the more important process, and hence must be longer and more minute, while in the investigations of an unknown science the analysis is the more important and valuable process.

In the general synthesis of Logic we shall also devote a chapter to the subject of Fallacies, and then consider some of the ways in which the syllogism is used, and the technical phrases which express these uses.

3. A Historical View of Logic.

This historical view of Logic has been placed after the study of the formal Logic, rather than before it, as is usual in most treatises, because we can appreciate a history only of that which we know, and we shall understand much better the causes of error and the obstacles to science which history gives us when we are beforehand aware of the true scope and relations of the particular science whose history is related. When we know what Logic is, its history is intelligible and interesting, and not otherwise.

For Logic is so intermingled, or rather entangled, with other kinds of philosophy in almost all of its principal epochs, that any one who should undertake to read of its adventures in history, without being able constantly to dissociate it from its companion sciences, would find it a useless and unprofitable task.

CHAPTER II.

ANALYTICAL VIEW OF LOGIC.

(11.) The Reasoning Process Analyzed.

To apply the method of analysis to the study of Logic as an art, we begin with the definition already laid down that Logic is the Art of Reasoning.

Reasoning consists in the combination of two known judgments to form a third, which is deduced from them. Reasoning, when expressed in language, is called *argument*.

The ultimate and simple form of argument, logically expressed, is *the syllogism*.* In a more extended sense, reasoning covers also the combination and succession of many arguments.

The *syllogism* is an argument consisting of three *propositions*, of which the first is called the *major premiss*, the second the *minor premiss*, and the third the *conclusion*. This is the usual order of the premisses, but the reasoning would be equally valid were they transposed.

> *Major premiss.* All A is B = All men are mortal.
> *Minor premiss.* All C is A = All Hindoos are men.
> *Conclusion.* Therefore all C is B = All Hindoos are mortal.

Each of these *propositions* consists of two *terms*, the *subject* and the *predicate;* and the verb uniting them is called the *copula*. Men reason to satisfy their own minds, to demonstrate truths, or to refute error, and, in so doing, they combine many of these syllogisms, thus forming *compound arguments*, which may always be analyzed into the simple arguments which compose them. In a simple syllogism, in many

---

* συν and λογιζομαι, more remotely λεγω.

cases, one or other of these premisses conveys a fact so well known that it may be taken for granted, and so it is suppressed, and thus is formed an *abridged argument*, called an *enthymeme*. For example:

> (*Minor premiss*) Cæsar was a man,
> Therefore Cæsar was mortal.

This is an enthymeme with the major premiss suppressed. This major premiss is, *All men are mortal*, which is taken for granted in the conclusion, where, because *Cæsar was a man*, it is affirmed that *he was mortal*. In every case, however, if the enthymene appear at all doubtful, the suppressed premiss may be written out, and the validity or invalidity of the argument thus determined. *Compound arguments*, instead of having each syllogism fully expressed, are usually formed of a number of enthymemes combined.

The groundwork of the syllogism is the *dictum of Aristotle*, or his universal test for Argument.

Without in this place entering even very briefly into the History of Logic—a history of experiment and error—it is interesting to know the time of its first decided manifestation, and the person to whom we owe it as a definite science. In that magnificent period when the school of Plato had prepared the mind of Greece for the coming of Aristotle, and the energy of Philip had opened the way for the conquests of Alexander, that system of Logic was formed, which, after having passed through the fiercest ordeals, has remained almost without change to our day. It has been indeed covered up, and to all appearance lost, in the times of European bigotry and ignorance; schoolmen and churchmen have alike assailed it; but, with the vital principle of truth, it has remained untouched by the ruinous hand of Time, amid exploded systems of Ethics, false speculations of Philosophy, and the cunning allegories of Heathen mythology. The Analytics of Aristotle form the cyclopædia of Logic in this age, as in all former periods.

THE DICTUM OF ARISTOTLE.

After many years of patient investigation Aristotle established the *"Dictum de omni et nullo,"* of which the first part, *de omni*, refers to all affirmative reasoning, and the second, *de nullo*, to all negative reasoning. Stated by the use of ordinary symbols it would be written as follows:

The Dictum of Aristotle.

De omni.
All A is B.
(1)   (2)
All or some   C is A.
(1)   (2)
Therefore all or some C is B.

De nullo.
No A is B.
(1)   (2)
All or some C is A.
(1)
Therefore no C is B, or some C
(2)
is not B.

Writing out the forms separately, we have—

De omni.

(1)
All A is B.
All C is A.
All C is B.

(2)
All A is B.
Some C is A.
Some C is B.

De nullo.

(3)
No A is B.
All C is A.
No C is B.

(4)
No A is B.
Some C is A.
Some C is not B.

Or, if stated by a geometrical notation, as all syllogisms may be stated:

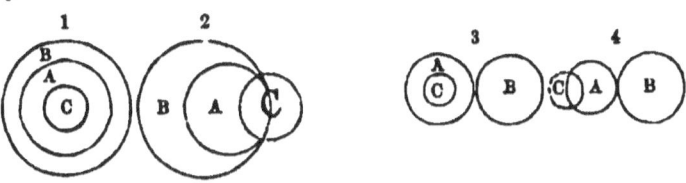

But to explain the dictum practically, it has been translated thus:

Whatever may be predicated of a whole class, may also be predicated of all or any of the individuals contained in that class.

To *predicate** means to *affirm* or *deny*.

Thus in the *dictum de omni*. In the major premiss we predicate or affirm *B* of the whole class *A*.

In the minor premiss we assert that all or some C is an individual or a number of individuals included under the class A.

And in the conclusion we predicate B of the individuals, as we did in the major premiss of the whole class to which they belong.

This simple dictum of Aristotle is the groundwork of the syllogism, and the syllogism is the universal principle of reasoning. It is sufficient in this place to state the fact; it will be proven hereafter. The propositions of which the syllogism is composed are further analyzed. A *proposition* consists of two *terms* and a *copula*, of which the *first* term is called the *subject*, the *last* the *predicate*, and the *connection between them* is the *copula*.

subj.	cop.	predic.
(Men)	(are)	(mortal)
subj.	cop.	pred:
(Men)	(are not)	(trees)

It has been said that the dictum of Aristotle is the groundwork of the syllogism, and that the syllogism is the universal principle of reasoning: it must be also remarked that every valid argument, no matter what may be its original form, may be put under the form of the syllogism, and to it in that form the dictum may be directly applied; and, on the other hand, if any argument cannot be reduced to this form, it is invalid. Thus this dictum forms not only the vehicle of correct reasoning, but is a sure test of error in Logic. We shall

* *Prædico—are.*

constantly recur, in considering every form of argument, to this test.

The reasons why in mathematical investigation we use letters, and in arithmetic numbers, are—first, to *expedite* and *simplify* the work, and secondly, to *generalize* it. For the same purposes we use symbols in Logic. If, for example, I write the syllogism

> All good men are happy;
> John is a good man,
> Therefore, John is happy,

I limit my argument entirely to the particular of *John being a good man* and *being happy*, whereas, if I write

> All A is B;
> C is A,
> Therefore, C is B,

I propose a general formula which will apply to many cases according to the subject and the matter of inquiry. It will be well for the student to frame particular examples under the general formula, and thus at once to fix the form in the mind and accustom himself to the practical applications of the system of Logic to particular cases.

Besides the dictum of Aristotle, to the form of which every valid argument may be reduced, there will be given hereafter a series of rules for detecting fallacy and for determining the validity of an argument when it is not exactly in this form, and, by means of these, the logical student may defend himself against the subtlest sophistry, holding Aristotle's dictum in reserve as a final test. Where one who is ignorant of Logic is obliged to use much effort and circumlocution to determine the validity or invalidity of an argument, and is in great danger of error in the process, the logician, at once and without inquiry into the subject-matter of discourse, applies his tests to the framework of the reasoning, and indicates infallibly the defect in the argument. And so deciding as to the validity or invalidity of the general formula as expressed by the symbolical letters A, B, C, he has once for all decided for each particular example which can fall under that formula.

In concluding this brief analysis of Logic, let us recapitulate. Logic is the Art of Reasoning. There is but a single universal principle of Reasoning. Reasoning here includes the consideration of terms, considered either as intuitions or conceptions, their combination by the judgment into propositions of various kinds, and the union of propositions into arguments as premisses and conclusion. All these processes are conducted in accordance with the laws of thought. The basis of reasoning is the dictum of Aristotle, and its simple form is the syllogism.

The syllogism is composed of two premisses and a conclusion; each of these is a proposition, and each proposition consists of three parts, two terms and a copula. It is now our purpose to examine these constituents of Logical formulæ in the inverse order, beginning with terms.

CHAPTER III.

A SYNTHESIS OF LOGIC.

(12.) Of certain Operations and States of the Mind in the Process of Argument.

IN proceeding to the synthesis of the reasoning process, we must first consider certain operations and states through which the mind passes in approaching an argument. Logicians have enumerated many which are so nearly related to each other that we may reduce them to three:

These are: 1st. *Apprehension;* 2d. *Judgment;* 3d. *Reasoning,* or *Ratiocination.* As a preparation for these in their order, *Attention* has been called the primary state. Attention is not a distinct faculty, but an act of will subordinate to intelligence—a general phenomenon of intelligence; but this is self-evident. *Apprehension* is a pure conception of an object, whether as perceived by the senses or otherwise presented to the mental consciousness. The idea or notion of the object is the fruit of this operation of the mind.

By the five senses of the body we have a knowledge of the world around us; the first step in obtaining this knowledge is sensation, *or the impression on the organ of sense;* sensation is conveyed in a mysterious, inexplicable manner, to the mind, to produce perception; and as soon as we have perceived the object by this union between the mind and the senses, the object is apprehended or taken hold of by the mind, and the idea is formed or an intelligent knowledge of it is produced.

Ideas are simple or complex.

A *Simple* idea is the notion of *one* object, or of several which bear no relation to each other; and this notion is expressed generally by one word, as *John, man, river;* or by many connected by conjunctions, *John and Peter, the man and the boy.*

A Complex idea is the notion we form of several objects which bear a relation to each other, as *a man walking, a bundle of rods.*

When an idea produced by an act of Apprehension is expressed in language it is called a *term.*

But, whereas certain *words*, which express terms, are equivocal or ambiguous, it must be observed that Logic deals only with general or abstract terms, and has nothing to do with their distinctness or indistinctness. It only takes for granted that a term is distinct and unambiguous. A Logical *term* is the expression in language of an idea obtained by *act of apprehension.*

2. JUDGMENT.

Judgment is that operation of the mind by which, if we have two *objects of apprehension* or *terms*, both known to us, we declare that they agree or disagree with each other. Thus, if I know who *"John"* is and what *"a hero"* is, I may declare that—

John is a hero,
Or that—*John is not a hero.*

Judgment is therefore of two kinds—*affirmative* when the two terms are declared *to agree*, and *negative* when they are declared *to disagree.*

An act of Judgment, when expressed in language, is called a *proposition.*

And here also it must be observed that Logic only takes cognizance of abstract propositions, which are expressed by logical formulæ, and has nothing to do with their truth or falsity; or rather, it takes for granted, indeed, that when a proposition is stated it is *true.*

For example, if the proposition be *A is B*, it is assumed by Logic that *A is in reality B*, and thus, if, when this general formula be translated into a particular proposition, it prove to be false, Logic is not responsible for the falsehood, nor for the error which finds its way into an argument by reason of the use of a false premiss. Much error has arisen through the mistake of supposing that Logic had to do with Language directly, and with the judgments expressed in language; but it is just such an error as would lead us to assign such values to the unknown quantities in any algebraic formula, such for instance as $y^2 - 2px = 0$, as would destroy the equation. Algebra presupposes the equation to be just, and develops only such values of x and y as will establish it. The Logical formula is as abstract and general as this, and Logical propositions are always assumed as true.

3. RATIOCINATION.

Ratiocination is that act of the mind by which, having two or more acts of judgment, or *propositions*, we pass to another or others founded upon them and growing out of their combination.

Thus, if we have the two propositions

> *All men are mortal,*
> *Cæsar was a man,*

we have, as an inference or fact implied in these two propositions, and deduced from their combination, the final proposition *Cæsar was mortal.*

An act of ratiocination, when expressed in language, is called an *argument;* and an argument, when reduced to its simple logical form, is called a *syllogism.* That simple logical form demands a certain order in the premisses and the conclusion.

If now we examine the syllogism

> *Major Premiss.* A is B = Men are mortal,
> *Minor Premiss.* C is A = Cæsar is a man,
> *Conclusion.* C is B = Cæsar is mortal,

we shall perceive that it consists of *three* propositions, which are called the major and minor premisses and the conclusion, and *three* terms represented by A, B and C, each term being used twice in the syllogism. The term which occurs in the major premiss and the conclusion (B) is called the *major* term; that which occurs in the minor premiss and the conclusion (C), the *minor* term, and that which is found in both premisses (A), the *middle* term. The major term is always the predicate of the conclusion, and the minor term the subject.

Extended ratiocination is conducted by the combination of many of these syllogisms or their conclusions, according to Logical laws.

CHAPTER IV.

(13.) Of Terms.

A TERM has been defined *an idea expressed in language*, and may be either *simple* or *complex*. As we shall see hereafter, two *terms* are connected in a proposition, and the name is derived from this fact, since they constitute the *termini* or boundaries of a proposition.

A *simple term* expresses a single object of apprehension, and is generally *one word*, as *man, house, field*.

A *complex term* is the expression of several objects of apprehension with the relation which they sustain to each other, as *a good boy, a horse running*.

It is evident that the term itself is arbitrary, and of use only to convey the apprehension to another, as in different languages the terms which express the same object of apprehension will be different words; thus we have the object we call *horse* expressed in French by the word *cheval*, and in Spanish by the word *cabállo*. Words, then, it must be remembered, are not acts of apprehension, but are arbitrary signs for expressing them.

But *language*, or the use of words, is necessary to the form of reasoning, as no reasoning can be applied and tested until it assumes the dress of language.

When a word is capable of being used alone as a term, it is said to be *Categorematic*,* and when it needs the assistance of other words to constitute with it a term, it is called *Syncategorematic*. Thus *man, horse, John*, are categorematic words; *here, gave, and*, are syncategorematic.

* Κατηγόρημα = something alleged or affirmed.

By a casual examination of the different parts of speech we shall find:

1st. *Of the noun:* That it is only categorematic when in the *nominative* case; the *possessive, man's,* requires another word denoting the thing possessed, and the *objective* a word which governs it.

2d. *Of the adjective:* That it is syncategorematic; for, although we say *John is good,* we understand *man* or *boy* after *good.*

3d. *Of the verb:* That it is, so to speak, more than categorematic, or *hypercategorematic,* since it contains often the *copula* and the *predicate:* as, *the man walks;* in this sentence *walks* is equivalent to *is walking,* in which *is* is the copula, and *walking* the predicate.

The infinitive mood is often in reality *not* a verb, but a noun in the nominative case. Thus the sentence *To die for one's country is happiness,* means *Death for one's country is happiness; To die* being fully expressed by *Death.*

4th. Of the remaining parts of speech we see at a glance that they are syncategorematic, and are only used in connection with other words to constitute terms. The word which has the form of the *present participle* is sometimes an *infinitive,* and sometimes *a noun;* we might substitute it in the last example given as a case of either. *Dying for one's country is happiness,* is equivalent to both the forms given.

(14.) Division of Simple Terms.

Simple terms are divided into *singular* and *common.*

A *singular* term is that which expresses a single individual, and is usually the *name* of a person, place, or thing; as *John, Philadelphia, the Delaware.*

A *common* term is that which expresses any individual or individuals of a whole class; as *a man, the men, an army.* To make a common term singular, we prefix the demonstrative pronoun *this* or *that,* as *this man, that river,* which is equivalent to stating the name of the man or river; as, *This man* is *John;* *That river* is the *Delaware.* *Common* terms stand for classes, and are sometimes called *appellative,* as giving name or appellation to many individuals.

They thus are of great aid to science, in that, when many common properties have been discovered in a great number of individuals, and their distinctive peculiarities have been discarded, they may all be called by one name, and that name will be a common term; when this is in view a common term is called, according to its comprehension, *genus* or *species.*

Common terms are further distinguished, *according to their matter,* into *abstract* and *concrete.*

An *abstract* term is an ideal word, expressing an abstract property capable of inherence in an object, and yet without reference to that object. Thus *hardness, length, beauty,* are abstract terms, which inhere in many objects, but do not indicate any particular one.

A *concrete* term is one which presents to the mind, at once, the property and the existence of the object in which it inheres. Thus *hard, long, beautiful,* are *concrete* terms, implying certain objects which are *hard, long,* or *beautiful.*

Concrete terms are also called *denotative* and *connotative,* because they *denote* the *abstract property,* while they *connote* or imply in their signification the *body* or *object* to which it belongs. Thus *hardness,* being an abstract term, is also an ideal noun; the mind rests upon the vague idea, because it indicates nothing farther; but when *hard* is mentioned we feel the right to ask, *what is hard?* the answer is—*stone.* Thus the concrete term *hard* has *denoted* the quality of hardness, and *connoted stone* as the object in which that quality inheres.

Terms are also divided into absolute and relative. An *absolute* term is one which does not refer to any other.

A *relative* term is one which refers to or implies another. Two terms which have a necessary relation to each other are called *correlatives.* Thus *father* and *son, king* and *subject, brother* and *sister,* are correlatives. Sometimes one term has several relations, or more than one correlative. Thus *nephew* implies *uncle* or *aunt*, and the *brotherhood* of *father* or *mother* with sister or brother.

(15.) Quality and Quantity of Terms.

Terms are further divided according to their *quantity* and *quality*.

The *quantity* of a term expresses how much of it is taken or considered.

The *quality* of a term is the mode or manner in which it expresses an idea of an object.

Quality is essential or accidental. An essential quality is that without which we cannot conceive of the existence of the object; such as *sense* and *intelligence* in man; *length*, *breadth*, and other dimensions in body.

An accidental quality is one which the object may have at one time and not at another; as *whiteness* in a wall; *health* to the body.

Terms are said to be *synonymous* under this division, when they express the *same* act of apprehension; but by common usage this exact meaning is departed from, and synonymous terms now mean those which express different shades of meaning; thus *happiness* and *felicity* are synonymous terms, and yet their etymology teaches us a difference in their meanings; the former attributing pleasure to luck or fortune, and the latter simply asserting a state of unalloyed pleasure.

Incompatible terms are those which cannot be used as predicates of the same subject at the same time: thus *hot* and *cold*; *asleep* and *awake*.

Positive terms are those which state the real existence of the objects they stand for. The opposite of these are *negative* terms, or those which deny the existence or assert the absence of certain objects or attributes.

There is a class of terms called *Privative*, which are often confounded with *negative* terms; but there is a real and im-

QUALITY AND QUANTITY OF TERMS. 43

portant difference between them. A *privative* term expresses that some quality or attribute usually belonging to the class is wanting in some individuals of that class: thus *dumb*, *idiotic*, are privative terms, since their very names call to the mind the fact that man generally is gifted with speech and reason, while *negative* terms denote the absence of a quantity or property which is not due to the subject.

Terms are divided according to their quantity into many distinct classes, expressing their number and dimensions.

Thus we have the common division of numeral and ordinal, as *twenty*, a *hundred*, *two*; *positive* (in its grammatical sense), *comparative* and *superlative* terms, as *good*, *better*, *best*.

That which is more truly a logical division is into *distributed* and *undistributed*: a *distributed* term being one the whole of which is considered, and an *undistributed* term one of which only a part is taken, this part being usually an indefinite part, expressed by such words as some, few, several, etc. *All men* is a distributed term, *some men* an undistributed term.

CHAPTER V.

OF THOSE OPERATIONS IN LOGIC WHICH RELATE TO TERMS.

(16.) Abstraction and Generalization.

COGNITIONS, INTUITIONS AND CONCEPTIONS.—A *cognition* is the impression which an object makes upon our mind, so that we *know* it.

An *intuition* is the knowledge or cognition we have of a single object, as *this house; the State house; John, the Hudson.* The mind receives an intuition, by simply attending to the object. This is a technical use of the word *intuition.*

A *conception* (*con* and *capere*) is a notion formed by gathering several objects into one, as *river, man, house.*

Conceptions are formed by the processes of abstraction and generalization.

ABSTRACTION consists in *drawing off and considering one or more of the properties of an object to the exclusion of the rest.* Thus we use *abstraction* when we observe the color and odor of the rose, disregarding its other characteristics. If we abstract the color and odor of one flower, then of another, and so of many, and finding these alike for all, call them all by one common name *Rose*, we are said to *generalize*. Abstraction aids us in passing from the confused and complex to the distinct—always dividing and simplifying: it is both *positive* and *negative*, considering one or more by the negation of others.

GENERALIZATION, then, consists in *disregarding the differences between many objects which are alike in certain properties,* only considering those which are alike and calling them by a common name—and thus it is that general and universal ideas are obtained.

We may abstract, it is evident, without performing the other process of generalizing, but we cannot generalize without first abstracting: in the general case, however, we abstract for the purpose of generalizing. It is by these two processes that we obtain *common terms*, or *the names of classes*. All these common terms are the result of higher or lower processes of generalization. Thus, by a low generalization, we obtain *tea-rose*, by a higher, *rose*, by a higher still, *flower*, and by one step farther, *vegetable*, etc. But common terms, as *classes*, are further distinguished into *species* and *genera;* and, as expressive of certain things *belonging to the species and genus*, they are also divided into the *differentia, property*, and *accident*. Some writers, in considering the *substance* of a term, have called the object for which it stands, the *essential* part or the *essence*.

A *class* denoted by a *common term* may be considered according to its *intension* or *extension*. By *intension* (also called *comprehension*) is meant, the inclusion of fewer objects with more specific differences; and by *extension*, the inclusion of a greater number of objects with fewer specific differences. Thus a species has more intension than its genus, the genus more extension than its species.

(17.) Species, Genus and Differentia.

A *species* is a class obtained by generalization, which includes only individuals or subordinate classes, and is itself included in a genus: as *an Arabian horse* is a *species* of *horse, horse* is a *species* of *quadruped; quadruped* is a *species* of *animal.* A *genus* is a class obtained by a higher generalization, which comprehends under it two or more species; as *animal* is the *genus* alike of *quadruped* and *biped, quadruped* is the *genus* of *horse, cow, deer*, etc., and *biped* the genus of *man*, etc.

It is evident that in one sense the species implies more than the genus; as, for instance, if *quadruped* be the *genus* and *horse* the *species*, *horse* will contain all the signification of *quadruped*, and also the distinctive signification of *horse* as to shape, size, habits, uses, etc.; which latter does not belong to *quadruped*.

For this reason the *species* is said to express the *whole essence* of the object, while the *genus* expresses only *a part of the essence*, and *that* the *material* part, or part common to all the species under that genus. Thus, *man* expresses the whole or complete essence of the animal so called, while *animal* expresses only the comprehensive or material part of the essence which only limits him to an animate existence.

The *differentia* of an object is the formal or distinguishing part of that object, and divides it from a class to which it does not belong; and when united with the *genus* or material part, or part common to all, forms *with it* the *species*, or whole essence. Thus, if *man* be the *species*, and *animal* the *genus*, *rational* would be the *differentia*, and we should have man (species) = rational animal (differentia) (genus); by which it appears that although the extension of the genus includes *this species* and *many others*, the *species really comprehends*, although in a different sense, more than the genus—namely, the genus and differentia—while the genus expresses only the material part, or that common to all. The genus has greater *extension*, i. e., extends to more classes and individuals; but the species has more *comprehension* or *intension*, i. e., includes the part expressed by the genus, besides the specific difference.

It is manifest that the differentia may be of three kinds: *generic*, as for instance the difference between *man* and *tree*; *specific*, as that between the different species, *horse* and *cow*; and *individual*, as between *Byron* and *Moore* as poets; but each becomes, in reference to the genus above, a specific difference

(18.) Property and Accident.

Thus, having shown the relations between the species, or the whole essence, the genus, and the differentia, parts of the essence, each of which is expressed by a common term, we come to consider those things which are or may be *joined* to the species or essence. They are divided as follows:

I. *Property*, which is joined universally to the essence, and thus must be asserted as belonging to every individual of the species; and, 2d. *Accident*, which is joined only contingently, that is, to one individual or *certain individuals* of the species, and not to the whole species.

Property is of two kinds: 1st. That which is *universal*, or belonging to every individual of the species, *but not peculiar to the species*, as *respiration*, which, although it *belongs to all men*, is *not confined to the species man*. 2d. That which is *universal and peculiar*, as *the power of intelligent speech*, which, while man as a species possesses it, is peculiar to man. Some writers have erred in enumerating a third kind, viz.: *peculiar, but not universal*, as, for example, *to be able to be a poet*. But this violates our definition, since, if it belong to some individuals and not to the species, it ceases to be a *property*, and becomes an *accident*.

II. *Accident is something joined contingently to the species, or belonging only to certain individuals of it.*

Accident is of two kinds, *separable* and *inseparable*. A *separable accident* is a circumstance which may be detached from the individual without affecting his identity or altering our general conception of him; as John is *walking* or is *lying down*; in which examples the *accidental* circumstance of *walking* or *lying down* is not a necessary part of the individual, but may be detached from him, so that we may still conceive of him as doing neither.

An *inseparable accident* is one which cannot be detached from the individual; as, *born in Philadelphia, born in* 1800.

It is by means of such inseparable accidents that a man is

described or his history written; but it must be remarked that this phraseology is rather convenient than exact, for as soon as the event which we call a separable accident occurs in the life of an individual, it really becomes *inseparable*. Thus, if John *walked* to the city on a certain day, or, being unwell afterwards, was *lying down* in consequence, we can no more detach these facts from his history than we can the event of his *being born* in a certain place and at a certain time; but as they are unimportant, we make no life-record of them.

Having now illustrated the meanings of *genus, species, essence, differentia, property* and *accident*, let us, for convenience and clearness of illustration, write out a sentence embodying all these uses of common terms, as a model by which the student will easily frame other examples for himself. This sentence will also embody the different processes of generalization.

(individual)　(species)　(differentia)　(genus)　(property universal but not peculiar)
John is a *Man* = *a rational animal, who breathes, has the*
(property universal and peculiar)　(separable accident)　(inseparable accident)
faculty of speech, is lying on the sofa, and was born in Philadelphia.

The logical name given to every common term representing a *genus, species, differentia, property, accident,* is *predicable;* viz., something which *may be predicated:* no other terms than these are *predicable*.

(19.) Of the Different Orders of Genera and Species.

A *summum genus*, or highest genus, is the highest class of all, and has no genus above it.

A term which expresses at once a *genus* and a *species* is called a *subaltern genus and species*. For example, *quadruped* is a *genus* of *horse* and a *species* of *animal*.

In the descending scale from the *summum genus*, the successive or inferior genus is called a *subaltern genus*.

DIFFERENT ORDERS OF GENERA AND SPECIES.

In the ascending scale from the lowest *species*, it is called the *subaltern species*.

When a *genus* is divided into its species, they are called *co-ordinate* or *cognate* species, to indicate that they are not *subordinate* to each other. Thus, if quadruped be divided into *horse, cow, lion*, as representing the *equine, feline* and *vaccine* races, these would be *cognate* species.

A species which contains beneath it *no other species*, but *only individuals*, is called an *infima* or *lowest species*. In any scientific investigation, however, ranging between any two limits, *although not absolutely the highest and lowest*, it is usual, *for convenience*, to call the highest limit named *summum genus*, and the lowest *infima species;* as though we should say, "Let A be the summum genus and C the infima species during this investigation." There are also in common use the phrases *proximum genus* and *remote genus*, the first of which means *the genus next above*, and the second *a genus farther removed* from the species in question. Thus, *quadruped* is the *proximum* and *animal* the *remote* genus of *horse.* It is necessary that the proximum genus should be the genus next above the species in question; but the remote genus may be any one farther removed, and not necessarily the *summum genus*, which is, of course, the *most remote*.

It must be observed that the use of a common term, as either *species, genus, differentia, property* or *accident*, is a relative use; and because it is used with one of these significations in one sentence, this does not deter us from using it with quite another meaning on another occasion. Thus if we take the word *red*, we shall find we can make it serve as each in turn.

The color *Red* is a *genus* under which as species are ranged *pink, scarlet, crimson, vermilion*, etc., the different kinds of *Red.*

Red is a *species* of the genus *color*, and ranges with white, blue, yellow, etc., as *cognate species*.

Red is a *differentia* of the "*Red rose,*" which distinguishes it from other roses. *Red* is a *property* of *blood;* and an *accident* of *a house, separable* if it be *painted red, inseparable* if it be built of *Red stone.* And thus in analyzing any sentence we must be careful to ascertain the real value of the common terms employed.

(20.) Realism and Nominalism.

While upon the subject of common terms, it is well to refer to the long-standing controversy between the *Realists* and the *Nominalists,* which, although it became strangely intermixed with theology and church polity, had its origin in the *significance of a common term.* It will be referred to more at length in the historical view. The *Realists* contended that every common term was *the name of something really existing*—that a genus and a species were *real things;* while the *Nominalists* believed that we obtained common terms merely to express a certain inadequate, undefined notion of one individual, which we apply to many, and that thus species and genera are mere names that have in nature no corresponding reality.

It would seem to be a trivial subject for controversy, but the more we examine it the more difficult and subtle it appears. Like many subtle controversies, it seems to be of little consequence in which way it could be decided; but it had, to the disputatious Greeks and the more disputatious Schoolmen, a charm on account of its subtlety, which its value could not secure to it.

Not to detain the student, let us state the true nature of the question, and solve the difficulty by saying, that genera and species are merely universal ideas, and as such exist only in the mind; that they are expressed by common terms, but that they have a real foundation in the individuals from which they have been acquired.

(21.) Definition of Terms.

Definition[*] is applied to *terms* in their logical use, and means *limiting them in such a manner as to distinguish them from all and any other terms.*

As much error arises from the indistinctness of terms, and the fact that different persons employ them in different meanings, just definitions which may bind both parties in a controversy are very important.

A definition is usually put in the form of a categorical proposition, of which the *subject* is *the term to be defined*, and the *predicate* is the definition proper. Thus in the example "*Man is a rational animal,*" the whole sentence is called *the definition*. This is not, however, strictly speaking, correct; as the predicate alone, "*rational animal,*" defines " man," as if in answer to the question " what is the definition of man?"

The first division of *definition* is into two kinds, *essential* and *accidental*. Essential definitions are further divided into *physical* and *logical*.

The second division of definition is into *nominal* and *real*. Before explaining the meaning of these divisions, we shall arrange them, for the sake of convenient reference, into a tabular statement.

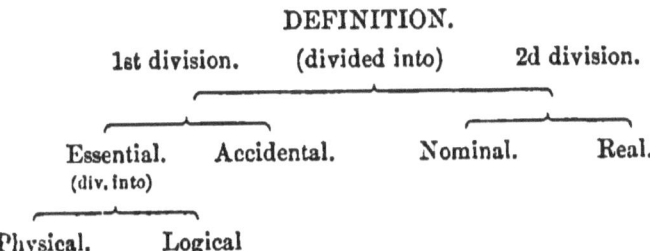

An *essential* definition is one which presents to us the *principal parts* of the *essence* of the thing defined; thus, a steamboat is "something consisting of hull, engine, wheel-houses, smoke-pipe, etc.;" or, again, it is " a vessel for water transportation propelled by steam." In each case the form of our

---

[*] *De* and finio, more remotely *finis*.

essential definition would be induced by the character of the person asking the definition, and according to the information he desired, but always in terms of the *essential parts* of the object for which the term stands. But it must be particularly observed that these principal or essential parts are of two kinds widely different from each other: *physical* parts or parts which are *actually separable by the hand*, and *Logical* parts, or those which *are only divisible by the mind*. To explain, a *physical essential definition* of a ship would be "an object which consists of hull, masts, cordage," etc., being the parts into which it may be physically divided; while the *logical parts* which would constitute a *logical essential definition* would be the *genus*, viz., "ocean vessel;" *and differentia*, viz., "of peculiar build;" which, as we have seen, when combined make up the species *ship*.

(species)   (genus)   (differentia)
A ship—is an ocean-vessel—of peculiar build.

A *logical* essential definition, then, in every case, consists of the *genus* and *differentia*. Logic is concerned with logical definitions alone, but examines the others to distinguish between them and logical definitions. And it is likewise true that the physical and logical definitions sometimes coincide, but this is of rare occurrence.

An *accidental definition*, or *description*, as it has been technically called, consists in presenting the circumstances belonging to an object, and these are its *property* or *accident;* as these are generally more descriptive of an animal or object than the *material part* or part common to all, which is the *genus*, or the *differentia* which distinguishes the species in question only from its co-ordinate species.

From what has been said before, it will appear that in *describing* a *species* we can only use *properties*, as *accidents* attach alone to individuals, while properties belong to *every individual* of a whole species; we should use, besides, properties which are *universal and peculiar*, since, as they belong to every individual of the species, and *to none out of it*, we thus find its own characteristics; whereas if we used the properties which were universal but not peculiar, we should only know characteristics which marked that species in common with others, and thus not define it. Thus if we should describe *man* as "a being who lived and breathed," this would not define or describe him *justly*. So, too, in describing an *individual*, as for instance in biographical notices, we should not use *separable* accidents which are not a permanent and necessary part of the *object*, but *inseparable* accidents which belong necessarily and permanently to it. For example, if we say " William was the Duke of Normandy who conquered England in 1066," we describe him by means of the inseparable accidents, viz., that he was Duke of Normandy, and that he conquered England.

(22.) Nominal and Real Definitions.

We come now to the second division of definitions, into *nominal* and *real*.

A *nominal definition* is one which gives the *meaning of the term* which is used as the *name* of the thing. In brief, it *defines the name*. Thus, "a *telescope* is an instrument for viewing distant bodies." "The *photograph* is a painting made by light on sensitive plates." "The *decalogue* is the table of the ten commandments."

A *real* definition analyzes and explains, not the *name* of the thing, but the thing itself; enumerating, besides, all its important characteristics and properties; thus, a real definition for a *telescope* would be a *treatise on the construction, powers, and uses of the instrument*, and a real definition of the *decalogue* would be given only *by reciting all its commandments*.

In the investigations of science it is evident that the aim is to obtain *real* definitions, and the fuller and more complete they are the greater their value; but since in Logic we have only to do with the *names* of things, and *not with their subject-matter*, or the conceptions which they convey to us, it is evident that we only need *nominal* definitions and not *real;* and indeed, with regard to matters of general information, a *nominal* definition will be sufficient to settle the grounds of a controversy; for while it is the *name* that indicates the individual or the class, the *definition explains the name.*

We may even, sometimes, provided both parties to an argument agree to do so, consider as a definition *something which is purely hypothetical,* but which still partakes of the nature of a definition; thus, for example, in an astronomical problem we say, "*let C be the sun's place in the heavens;*" or in any case, for purposes of illustration, "*let so and so be so and so.*" This form of definition is purely relative; for although, in reality, C is not the sun's place, *it is so relatively to the other points* on the diagram.

It must also be observed that it is not necessary to the *justness of a definition* that it should refer to *real things,* as, for example, we define an *unicorn* to be "*a fabled animal, having but one horn,*" and a *phœnix* to be "*a bird fabled to live without a mate and to rise from its own ashes.*"

(23.) Rules for Definition.

So important has the subject of definition been considered, that Logicians have laid down three rules for it, to which, if we adhere, we shall insure just and adequate definitions.

1st. The definition must give to the mind a clearer conception than the name of the thing defined, or it will be useless. The *clearness* of a definition is opposed by negative attributes; thus to define *man* as *not a quadruped* would be unsatisfactory in this respect.

In most of the arts and sciences this consists in putting a technicality into plain language, for those who are uninitiated; but if I am asked to define *cow,* a word understood by every one, and say that cow is a *ruminant quadruped,* I violate the rule. In the nomenclature of science many technical terms give, in one word, what it would require much circumlocution to express in common words. Accompanying this rule there is the caution that the character of the definition should depend upon *the subject* and *the persons addressed.*

2d. The definition must be *adequate;* that is, neither include other things than those necessary to define, nor exclude any *necessary* explanation of the thing defined.

Thus, if I define *bird* to be "*an animal that moves in the air by means of wings,*" I am too extensive in my definition; as that would include other animals than birds, as bats, flying fish, etc.; and if I define it to be "*a feathered animal that sings,*" that would be too narrow, as some birds do not sing.

3d. The third rule is rather a caution which grows out of the other two than a rule like them. It is, that *the words used* in a definition *should be sufficient* and of the proper kind to define the thing.

If we use too many words, we confuse the meaning and are liable to *tautology;* if too few, we are liable to obscurity. Thus, to say that "*a square is a four-sided figure with equal sides,*" would be true but not definite, as there may be drawn other parallelograms not right angled, with equal sides. If we say "*a parallelogram is a four-sided figure whose opposite sides are equal and parallel,*" we use too many words, as the equality of the sides implies the parallelism, and *vice versa.*

In the first case we err, because we do not exclude, in our definition of the square, all other figures; in the second, because we allow it to be supposed that there are four-sided figures whose opposite sides are equal and not parallel. Under the head of tautology comes what is called Defining in a Circle; *i. e.,* by using the term to be defined in the definition. *Right is man's power to do or not to do. Law is a legal ordinance; evil is that which is not good.*

The examples taken are broader and more apparent than those in which faulty definitions are generally used, but they render the error more obvious, and indicate to us the character of the danger to be avoided.

If we would see the practical necessity of definitions, we need but consider a few of the vague and inexact terms which we use in our ordinary speech, and which it seems a prevailing fashion to distort in their meanings. We shall recur to this subject under the general title of "Verbal Fallacies," but may now give a few illustrations of the value of exact definitions. Take for example such words as Necessity and Necessary, which may mean either an accordance with the invariable law of God, or an obedience to the blind decree of fate, according to the belief or skepticism of him who uses them. In its political sense, the adjective *necessary* has been said to be capable of certain degrees of comparison, as in the argument urged in favor of the Bank of the United States,* in speaking of the means *necessary* for carrying out the provisions of the Constitution, it was asserted that they may be classed under the three categories of *necessary, very necessary*, and *absolutely and indispensably necessary*. So also in religion, certain things are said to be *generally necessary* to salvation, while others are said to be *absolutely necessary*. Thus the technical sense of the word is entirely lost, as that refers to an *absolute* condition, *which cannot but be*, or *cannot be otherwise*, and therefore does not admit of comparison. Or if we would see a strange, conglomerate example of indefinite and erroneous terms, demanding a clear definition, take the war-cry of the French revolutionists, "*Liberty, Equality, Fraternity;*" no one word of which can express to the people a distinct idea, or will bear the test of a clear definition.

It has been a custom in nominal definitions to define one term by means of its synonym, borrowed from another language. Although our language is, in its structure and the great majority of its principal words, Anglo-Saxon, still the large number of French and Latin words which have been brought into it have formed terms synonymous with the original Saxon;

---

* Kent's Commentaries, vol. i., Lect. 12.

but, when they had become naturalized, as we had no use for two words *exactly* synonymous, wisdom suggested that they should exhibit shades of difference in meaning, which did not originally belong to them · so that few if any words are justly defined by their synonyms. Besides, as a *similar idea* among any two people would have its differences drawn from their own peculiarities of clime, and race, and manner of life and government, the synonyms when brought into the language would often express great differences at once, and without any effort on our part to cause them to do so. As a remarkable instance of this, let us see how very wrong it would be to define our English word *freedom* by its synonym *liberty*, which comes to us from the Latin; and yet, how many confound the two. Indeed these are historic words, and give us an insight into the times of their birth, wonderfully illustrative of the people and countries from which they came. *Freedom* is the personal, individual independence and right of every man, his *free doom;* i. e., free province or jurisdiction from his birth. Coming as it does from the Teutonic element in our language, it tells us of the free and independent Germans, who, by their own valor, overturned the great fabric of the Roman empire. They were men of the forest and mountain, inhabiting no cities—there were none in Germany till after the eighth century—but only roving where were the lordliest spoils, and claiming them as the reward of their personal *freedom*. On the other hand, *liberty* tells us of the Roman cities, of the sway of the Roman empire, and of Roman licentiousness; of a form of manumission, implying slavery; individuality merged into citizenship. To be a Roman citizen was to have attained the post of honor, open to all advancement in diplomacy and war. Nor is the spirit belonging to these words yet lost. While we cling like good citizens to our *liberty*, vouchsafed to us by the constitution of the country as Americans, we much more desire to keep well guarded that *freedom* of opinion, of speech, of action, which is our indefeasible right as men.

In view of the importance of just definitions, let us undertake no controversy or expression of opinion involving a vague and indistinct term, without demanding a definition, and agreeing to use it during the discussion.

(24.) Division.

It is of great importance in the consideration of common terms which stand for classes, that we should be able to divide them into all their several parts or significates. An *individual*, as its name indicates,* is incapable of logical division. It is only a *species* or *genus*—*i. e.*, a *class*, in more general language—which can be so divided.

Division is of two kinds, *physical* and *logical;* to these some writers add, improperly, *numerical division*.

Physical division, also called *partition*, is the actual separation of the physical parts of which a thing is composed. It is evident that an *individual* is capable of *physical division;* thus, an individual *tree*, as a *certain oak*, may be divided into *trunk, branches*, and these further subdivided into *bark, heart, leaves*, etc.; an individual *man*, as *John*, may be physically divided into *head, arms, trunk, legs*, etc. With this kind of division Logic has directly nothing to do.

Logical division, which takes place in the mind only and is only applied to classes, consists in separating a *genus* into its different *species;* and a *species* into the *individuals* composing it; and this in regular order from the *summum* genus to the *infima* species. Thus, the genus *tree* would be logically divided into *oak, maple, hemlock, fir, pine, elm*, etc.; and the species *oak*, into *red oak, white oak, live oak, scrub oak*, etc.; and each of these again into the individual trees comprising its *kind*.

It will be evident that in a just division, each one of the parts—denoting a species—will be *less* than the whole number which make up the *genus;* or any one of the parts— denoting an individual—will be less than the whole number

---

* *In* and *dividuus*, from *divido*, to divide.

which make up the species; or, as a test of the correctness of the division, we must be able to predicate the *summum genus* of any one of the parts.

If, for example, we have assumed *tree* to be the summum genus, we must be able to predicate *tree* of *oak*, or *live oak*, or any individual live oak.

It is evident that the same term may be *logically divided*, *according to race*, into *Caucasians, Malays*, etc.; *according to creeds*, into *Buddhists, Jews, Mohammedans, Christians*, etc.; *according to nation*, into *Americans, English, French*, etc. These cross-divisions must not be mingled or confounded; for example, to divide *man* into *Caucasians, Mohammedans, Americans*, etc., would be false and useless division.

The principle of division is best illustrated by a scheme, or inverted tree, in which are arranged *clearly, symmetrically*, and without *arbitrariness*, the different parts of the division.

SCHEME OF DIVISION.—SUMMUM GENUS.

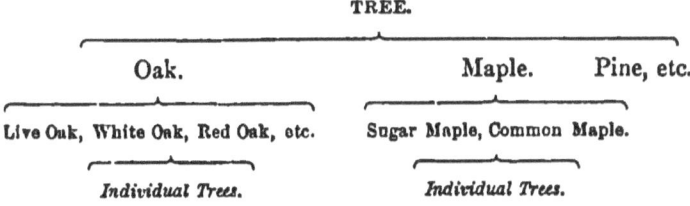

It may be well to observe particularly an auxiliary phrase, *according to*, which we use to keep us from a simple but dangerous error; *i. e.*, every division should be governed by one and a single principle. Man is divided not *into* races, creeds, nations, etc., but *according to* these, into various parts, thus:

SUMMUM GENUS.—MANKIND DIVIDED ACCORDING TO

```
                Race.            Creed.            Nation.
    Caucasian, Malay, etc.  Jews, Christians, Mohammedans.  English, French, German, etc.
```

It is evident that all the co-ordinate species must be on the same line or platform, that is, they must hold the same relative position to the *summum genus*. We must be careful to omit no *subaltern genus*; and we must place each *subaltern genus* in its own relative grade. Thus, if we should place *oak* properly, in the division of *tree*, but should pass immediately from the genus *tree* to the species *sugar maple*, thus leaving out the species *maple*, co-ordinate to *oak*, we should make an unequal and undue division. This would be placing one of the *co-ordinate* species on the same level with one *subordinate* to it. In other words generic, specific and individual differences must determine the systematic arrangement. To sum up:

I. The species constituting the genus must exclude one another.

II. All the species taken together must be equal to the genus divided.

III. The division must be made according to one single principle.

From what has been said, it will be seen that the process of Division is exactly the opposite of Generalization.

As in Generalization we *disregarded the differences* between many *individuals*, or between many species, and considered only the properties they had in common, that we might constitute them respectively *species* and *genus*, *calling them by a common name*, so in Division we take the *genus* thus obtained and *add to it the several differences* which we had removed in Generalization, and which distinguish its parts, that we may call the parts thus enumerated by separate names.

The two inverse processes of *generalization* and *division* may be plainly illustrated by a scheme or double tree; and this may be made as full as we please: thus, from individual trees we may generalize to the genus *tree;* or, from *trees* and *shrubs* and other kinds of vegetation, we may generalize to the sum mum genus *vegetable*. The division will be of the exact species, etc., but in the inverse order.

SCHEME OF GENERALIZATION AND DIVISION.

Individual Trees.	*Individual Trees.*	*Individual Trees.*
Live Oak, Red Oak, etc.	Sugar Maple, Common Maple, etc.	White Pine, Yellow Pine, etc.
Oak.	Maple.	Pine.

TREE.

Oak.	Maple.	Pine.
Live Oak, Red Oak, etc.	Sugar Maple, Common Maple, etc.	White Pine, Yellow Pine, etc.
Individual Trees.	*Individual Trees.*	*Individual Trees.*

What has been called *mathematical* or *numerical* division is in reality but a form of physical division; thus, I *divide* a loaf into *slices*, or an apple into *pieces*, *physically*, with or without regard to the equality of the pieces, or their sizes relatively to each other. If this equality or relation be observed, it may be called numerical division, but it is only an exact form of physical division; as a *half*, a *third*, *ten* times as great, etc., etc.

By a comparison of the subjects of *Division* and *Definition*, it will be seen that *division* is, after all, but a systematic and practical kind of *definition*, since there can be no better way to illustrate the meaning of *tree* than logically to divide it, before our eyes, into all its species down to individual trees.

It will be readily seen that the nature of the logical division of terms will depend much upon the science in which they are used, and the principle *according to* which they are to be classified. Thus, an *ethnologist* would divide *mankind* according to *races*, a *theologian* according to *creeds*, and a *statesman* according to *nation*. The *principle* of all the divisions would be the same, while the resulting cross-divisions, as we have seen, will be widely different.

(25.) Recapitulation.

It will be well to recapitulate briefly what has been said upon the subject of terms, and the various operations which concern them. We have shown—

1st. That a term is the expression of an object of apprehension, and have explained the different kinds of terms, according to a regular division.

2d. That common terms are obtained by the processes of Abstraction and Generalization.

3d. The distinction between *genera, species* and *individuals,* etc.

4th. The *Definition* of terms, and just rules for *definition.*

5th. *Division* of terms, with the difference between *physical* and *logical* division, and special consideration of the latter.

The next step will be to combine these terms into propositions; that is, from our knowledge of two of them to assert their agreement or disagreement.

CHAPTER VI.

(26.) Propositions.

A *proposition** is an *act of judgment expressed in language*, and consists of three parts, a *subject*, a *predicate* and a *copula ;* the subject and the predicate are called the *terms* or *extremes* of the proposition.

The *subject*, in the due order, is placed first, and is that of which something is predicated, *i. e.*, affirmed or denied.

The *predicate* is that which is affirmed or denied of the subject.

The *copula* is always, in categorical propositions, the uniting word which expresses the agreement or disagreement between the subject and predicate, and is always some part of the verb *to be*. When the copula is affirmative, *agreement* is expressed; when negative, *disagreement.*

subj. cop. pred.   subj.   cop.   pred.
A   is   B = (Cæsar) is (a tyrant).

subj. cop. pred.   subj.   cop.   pred.
A (is not) B = (Cæsar) (is not) (a tyrant).

The *negative* particle, it must be observed, *is always a part of the copula.*

What appear, in our ordinary speech, to be simple propositions are sometimes inverted or elliptical forms of expression, which must be put into simple logical form before they can be considered as propositions.

Thus we say "I hope to see you," "I desire to remain ;" and in these cases the subject is really placed last; the true meaning being

* From *propono*, something proposed or set forth for our acceptance.

subj.   cop.   prod.

(*To see you*) is (*the thing which I hope, or my hope*).

As an example of another form of inversion, we have that which springs from the constant use of the neuter pronoun *it*

Thus, in ordinary language, we say, "It is true that I think so." The true logical form may be given thus:

subj.   cop.   pred.

(That I think so) is (a true thing).

Many writers have denied that there is such a thing as a negative judgment, and consequently that any negation attaches to the copula; for they say that the proposition *John is not happy* is equivalent to *John is unhappy*, which transfers the negation from the copula to the predicate; but this is a quibble about words, as there are propositions in which the negation cannot be thus destroyed, and such is the case with far the greater number. The positive term is generally *limited* and intelligible, the negative *unlimited* and indefinite; thus, *man* is a term which we can grasp, but *not man* includes all the universe beside.

Of the Copula.—The copula may be always reduced to the present tense of the indicative mood of the verb *to be*, and consequently expresses neither *past* nor *future* time. Thus, "Cæsar *was* the conqueror of Gaul," is equivalent to "Cæsar *is* the historic personage who conquered Gaul." "I *shall be* glad to see you," is the same as "I *am* the person who will be glad to see you," etc.; but as this reduction is in general unnecessary, we agree to call those propositions which are expressed in time other than the present. Very often the copula and predicate are expressed together in one word, as "The sun shines;" here the word *shines* may be resolved into *is shining*, in which *is* is the *copula*, and *shining* the *predicate*. And sometimes, in other languages, as the Latin or Greek, a proposition is conveyed in one single word, as *amo*, I love or I am loving, τυπτω, I am striking; but in every case, a proposition may easily be placed in such a form that the subject, predicate, and copula are distinctly stated.

But this definition of a proposition, as a sentence consisting of a *subject, predicate* and *copula*, is evidently a *physical* definition, and is not sufficient for our purpose. The logical definition of a *proposition* is "*a sentence which affirms or denies;*" here *proposition* is the *species, sentence* the *genus*, and *which affirms or denies* is the *differentia*, or statement of the difference between this kind of sentence and all others. The word *proposition* not having in its etymology this strict meaning, it is very loosely used to express almost every kind of sentence. We must be careful, in Logic, to limit it to the definition just given. Hence, we should say that a categorical proposition, in its grammatical sense, implies the *indicative* mood, since absolute affirmation or denial is expressed only by that mood. Thus are excluded, the *imperative* mood or all *commands*, the *subjunctive* mood or all *hypotheses*, the *infinitive* mood, which, as its name indicates, is not a *finite, uniting* verb, but only a *verbal noun*.

If we examine these moods a little more in detail we shall find, first, that even in the indicative mood, *questions*, or the interrogative form of that mood, are excluded, for the use of a question implies that one of the parts of the proposition is wanting, and that we depend upon the answer to supply it. Thus the first and simplest form of the question is

$$Is\ A\ B? = Is\ man\ mortal?$$

If the answer be affirmative, then we have a right to the copula *is*, which before was wanting, and may write

$$A\ is\ B = Man\ is\ mortal.$$

Another form of the question is "What is A?" or "What is B?" the answer to which will supply us with the *predicate* and *subject* respectively. With regard to the *subjunctive* mood there are, it must be observed, propositions which assume that form and which are called *hypothetical*, and they come under the class of *compound* propositions, as

If A is B, C is D.

In almost every case the *hypothesis* is stated in the *indicative* rather than the subjunctive mood; thus,

If A *is* B, C *is* D; rather than in the form:

If A *be* B, C *will be* D.

Of the infinitive mood it may be observed that there are various forms; thus, *to ride is pleasant*, may be rendered by *riding is pleasant; horseback exercise is pleasant;* plainly showing that with the *verbal form* there is a *substantive value.*

(27.) **Propositions Divided into Simple and Compound.**

If, now, we proceed to consider first the *substance* of *propositions*, we shall find them divided according to their *substance* into *simple* and *compound*.

A *simple* proposition is one which has but *one subject* and *predicate*, united by the copula *is* or *is not*. Simple propositions are also called *categorical*, that is, there is simply affirmed or denied an agreement between the subject and predicate.

A *compound* proposition is one which has more than one subject or more than one predicate, and may be resolved into two or more simple propositions; as *The Delaware and the Schuylkill are rivers in Pennsylvania.* Compound propositions are further divided according to their substance into *categorical, modal, conditional, causal* and *disjunctive.*

A *compound categorical* proposition, like a *simple categorical*, affirms or denies the predicate *simply* and *certainly* of the subject; thus,

Alexander, Cæsar and Napoleon were ambitious of military glory.

A *modal* proposition is one in which the mode or manner of agreement or disagreement between the subject and predicate is stated, as *Cæsar conquered Pompey by unfair means.*

A *conditional* proposition consists of two simple categoricals united by the conjunction *if;* thus,

If A is B, C is D.

It is usual, for convenience, to place the conjunction first; the first categorical—A is B—is then called the *antecedent,* and the other—C is D—the *consequent.*

A *causal* proposition is one in which the *reason of the truth of* a simple proposition is stated; thus,

Because A is B, C is D.

A *disjunctive* proposition is one in which one of two or more simple propositions is asserted to be true; thus,

Either A is B, or C is D.

This is done by the use of the conjunctions *either* and *or.*

Propositions are still further divided according to two of Aristotle's categories which will be considered hereafter; *i. e.,* according to their *quantity* and *quality.* In simple language, *Quantity* considers of how much of the subject the predicate is affirmed or denied; as, *some* or *all* A is B.

And *Quality* regards the kind or manner of that predication, *i. e.,* whether it be affirmative or negative; whether A *is* or *is not* B.

(28.) Quantity and Quality of Propositions.

The *quantity* of a proposition is determined by the amount or portion of its subject which we consider. If we assert that the predicate agrees or disagrees with the *whole* subject, that is, all the significates which come under the term, the proposition is said to be *universal;* thus,

All men are mortal, no men are trees,

are universal propositions, because the whole of the subject is considered. But if we assert the predicate to agree or to disagree with only a *part* of the subject, the proposition is called *particular.*

Some men are brave, few men are good, many men are not prudent, are examples of particular propositions.

The *quality* of propositions we shall find also to be of two kinds—the *quality* of the *subject-matter* and the *quality* of the *expression*. Propositions are divided, according to the quality of the *subject-matter*, into *true* and *false*, and, according to the *form of expression*, into *affirmative* and *negative*.

It is evident that with the quality of the *subject-matter* Logic has *directly* nothing to do; for, since the logical form of a proposition is *A is B*, it is *taken for granted*, as we have already seen, that this statement is true, and that from the very form it assumes. With the subtleties of statements Logic is not concerned. Taking for granted the truth of a proposition, it makes use of it properly. Whatever falsity lies in it will pervade the argument, but this will not be the fault of Logic. In Logic the quality of the subject-matter is *accidental* and not *essential*.

The *essential quality* of propositions in Logic is, then, the *quality* of the *expression;* and this quality is made, as before shown, to depend upon the copula. If the *copula* is *affirmative,* the proposition is called *affirmative;* as

>All A is B.
>Some A is B.

If the copula is negative, the proposition is said to be negative; as

>No A is B.
>Some A is not B.

To mark these divisions according to *quantity* and *quality*, and to simplify the future operations in which they are used to frame arguments, we employ letters as symbols. Since every proposition must be *universal* or *particular*, and at the same time *affirmative* or *negative*, there are *four*, and only four, classes of simple categorical propositions, which we represent by the following symbols:

QUANTITY AND QUALITY OF PROPOSITIONS. 69

Universal affirmative; as *All X is Y*, by   *A*.
Universal negative; as *No X is Y*, by   *E*.
Particular affirmative; as *Some X is Y*, by *I*.
Particular negative; as *Some X is not Y*, by *O*.

The sign of a *universal* proposition is the same as that of a *distributed term;* i. e., the prefix *all* or *every* for the universal *affirmative*, and *no* for a universal *negative*.

And here it must be particularly observed that the universal negative is only correctly written when in the form *no* A is B. It might at first sight seem that this is equivalent to *all A is not B;* but it is not so, although often meant to be so; thus, *all soldiers are not cruel* has a very different meaning from *no soldiers are cruel.* The first is not, indeed, a *universal* proposition, as it appears to be, but a *particular*, implying that *some soldiers are cruel*, while *some are not.*

The translators of our English Bible have, in a few instances, made use of this form improperly to express a universal. Thus, the Hebrew text of the Psalms expresses with regard to the wicked: "All his thoughts are 'there is no God;'" while the translators have it, "God is not in all his thoughts." The meaning of the translators in this is evidently, "God is not in *any* of his thoughts."

The *sign* of a *particular* proposition is the same as that of an *undistributed term,* i. e., the prefix *some, few, several, many*, and like words, indicating a *part* only of a *whole*, for *particular affirmative* propositions; and the same prefix, with a *negative copula*, for *particular negative*.

But it constantly happens that a proposition has no prefix, and we are then thrown upon our knowledge of the *subject-matter* of the proposition, to determine whether it be universal or particular. Such propositions as have no prefix to denote their quantity are called *indefinite* propositions, which Logic alone will not enable us to understand. We must then look to their meaning, and thus find out what prefix is their due For example, *men are artists.*

By examining the matter of this, we find that only *some men are artists*, and then, making the proper prefix, we declare the proposition to be *particular*.

Birds fly. This is true of birds universally, and we have the right to prefix the sign *all*, which denotes it a universal proposition.

A *Singular proposition* is one which has for its subject a singular term; as

>Alexander was a conqueror.
>Cæsar was ambitious.

It would seem, at a first consideration of the quantity of these propositions, that they were *particular*, but this is erroneous; they are evidently universal; since when I assert that *Alexander was a conqueror*, I mean the *whole* of Alexander, or Alexander *taken in his fullest extension*.

As a general rule, then, *singular* propositions are universal. There are many other divisions of propositions which are curious rather than useful distinctions. The above are all those necessary to a comprehension of the logical processes which follow.

(29.) **Of the Distribution of Terms in Propositions.**

Having treated of the quantity and quality of propositions, and observing that, as we have already seen, these propositions are to be hereafter used in the framing of syllogisms, we come to consider the *distribution of terms* in propositions, and to establish rules for this distribution. If we examine the four categorical propositions, with their geometrical notations—

Affirm. $A.$ $\begin{cases} \text{All X is Y.} \\ \text{Some X is Y.} \end{cases}$ Neg. $E.$ $\begin{cases} \text{No X is Y.} \\ \text{Some X is not Y.} \end{cases}$

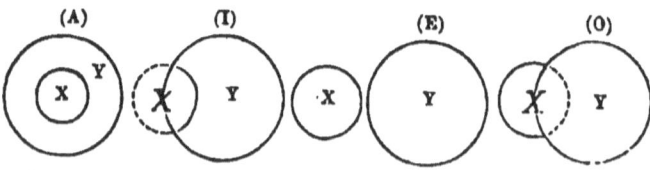

first with reference to their *subjects*, it will be evident that in

A and *E* the *whole* of the subject being considered, the subject is *distributed*, as is also indicated by the prefixes *All* and *No*. It will be equally evident that in *I* and *O* the subject is *undistributed*, a portion only being taken, as is indicated by the prefix *Some*.

The rule deduced then, as far as the *subjects* are concerned, is very simple; it is, that

All universal propositions distribute the subject. No particulars distribute the subject.

But since the predicates in these propositions have no such prefixes, how are we to determine whether they are distributed or undistributed? By an examination of the relation existing between the subject and predicate in each case, we shall see that the distribution of the subject by no means implies that of the predicate.

If we assert, 1st, that *All X is Y,* we do not assert that other things likewise may not be contained in Y; for though *All X is Y, All W may be Y; All Z may be Y,* etc.; or, to illustrate by a geometrical figure, we have

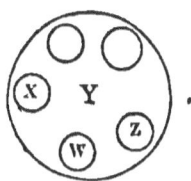

showing space enough for other things besides X to be contained in Y. Hence, it is evident that the whole of Y is not considered in the proposition *all X is Y,* or that Y, the predicate, is not distributed in a universal affirmative proposition.

Again, if we take the proposition *some X is Y,* the same reasoning will apply, since many other things may be Y, besides this *some X;* as illustrated in the figure

Likewise then we see that the whole of Y is not taken in this case, or that the predicate of a particular affirmative proposition is not distributed.

Thus far, then, we have found it true of *affirmative* propositions, *whether they be universal or particular*, that they *do not distribute the predicate.*

If, now, we consider the universal negative, *no X is Y*, we shall find that we must consider *the whole of X*, and *the whole of Y*, before we can assert that no part of one belongs to any part of the other; thus

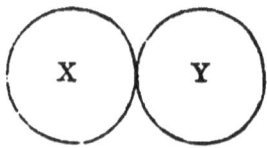

We have already seen that the subject *X* is distributed, and it thus appears that in a *universal negative proposition the predicate also is distributed.* The whole of the subject is brought in contact with the whole of the predicate, or we could not entirely deny their agreement. It remains now to consider only the predicate of a particular negative, *some X is not Y*. The same reasoning applies here as in the last case; or we must know and consider the whole of *Y*, before we can assert that no part of it belongs to the *some X* in question.

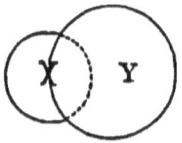

It therefore appears that the predicate of a particular negative proposition is distributed.

If we collect together these four results, we shall thus establish two rules:

1st. The subjects of universal propositions, and not of particulars, are distributed.

2d. The predicate of negative propositions, and not of affirmative, are distributed.

Or, all universals distribute the subject, and all negatives the predicate.

It may be well, for the sake of convenient reference, to arrange the quantity and quality of propositions, and the distribution of the terms, in a tabular form, so that it may be referred to until it be fixed in the mind of the student.

Four Classes of Categorical Propositions.	Subject.	Predicate.	Simple Form.
A. Universal affirmative.	Distributed.	Undistributed.	All X is Y.
E. Universal negative.	Distributed.	Distributed.	No X is Y.
I. Particular affirmative.	Undistributed.	Undistributed.	Some X is Y.
O. Particular negative.	Undistributed.	Distributed.	Some X is not Y.

There is a logical process which is passed upon propositions and upon propositions only, and this process has in view the use which we make of propositions in the framing of arguments. It is called *Conversion*. We cannot convert a *term*, nor is it proper to speak technically, as some writers have done, of the conversion of *arguments*.

(30.) Conversion.

Conversion consists in transposing the terms of a proposition in such a manner as to place the subject for the predicate, and the predicate for the subject. Thus, having the proposition *A is B*, we *convert it* into *B is A*. When no other change than this is made, the conversion is called *simple* conversion; but by an examination of the four forms of categorical propositions, it will be evident that they cannot all be simply converted, and retain in the converted proposition or *converse* the truth of *the original proposition* or *exposita*. As a simple example of this: having the proposition

All men are mortal,

we cannot write the *converse*,

All mortals are men.

No other conversion is allowed in Logic than that which is called *illative*,* or that in which we may infer the truth of the *converse* from the truth of the *exposita*.

To simplify this, let us convert each of these propositions in turn.

1. (A.) *All X is Y = All men are mortals.*

It is evident, as we have already seen, that we cannot convert this proposition *simply*, for we cannot read

All Y is X = All mortals are men,

since Y (or *mortals*) includes many other races besides *men*.

We, therefore, limit the quantity of the proposition from *universal* to *particular*, so that Y, which was *undistributed in the original proposition, may remain so in the converse*. Expressing, then, this non-distribution of Y by the prefix *some*, we shall have as the converse

Some Y is X = Some mortals are men.

From the nature of the process, this form of illative conversion is called *conversion by limitation.*†

From this we see that the *converse* of a *universal affirmative* is a *particular affirmative*, or A becomes, when converted, I. If we examine the universal negative,

2. (E.) *No X is Y = No men are trees,*

we shall see that as X and Y are taken in their whole extension, or are *distributed*, we may here convert *simply*, and read

No Y is X = No trees are men.

The *converse* of a *universal negative* is a *universal negative.* So, likewise, in the particular affirmative,

3. (I.) *Some X is Y = Some men are cruel,*

we shall find that neither subject nor predicate is taken in its

---

* *In* and *fero* (*latum*).

† The Latin name employed by logicians, for this kind of conversion, is *conversio per accidens*.

full extent or distributed, and that we may, therefore, convert simply:

Some Y is X = Some cruel (beings) are men.

The converse of a particular affirmative remains a particular affirmative. There remains only the particular negative to be considered.

4. (O) *Some X is not Y = Some quadrupeds are not horses.*

This proposition presents a special difficulty. We cannot convert it simply as in the cases of E and I; for we should then have X, which is *undistributed* in the *exposita*, *distributed* in the *converse;* thus we would have the absurdity

Some Y is not X = Some horses are not quadrupeds.

Nor can we invert the process of conversion by limitation as in the case of A (1), and pass back from particular to universal, as

All Y is not X = All horses are not quadrupeds.

To overcome this difficulty we detach the negative particle *not* in the original proposition from *the copula*, and attach it to the predicate; thus, instead of the usual form *some X is not Y*, we read,

Some X is (not Y) = Some quadrupeds are (not horses),

and then it is evident that for all logical purposes, the proposition ceases to be *O* or *particular negative*, and becomes I or *particular affirmative*, since for (*not Y*) we might place any other symbol, as Z, and convert by simple conversion. But without this trouble, if we convert, we shall have

Some (not Y) is X = Some (not horses) are quadrupeds,

or, in our ordinary language, to complete the sense,

Some (beings which are) not horses are quadrupeds.

This is called conversion by *contraposition* or by *negation*.

We arrive by this process at a rule for illative conversion, which is, that *No term must be distributed in the converse which was undistributed in the exposita.*

By arranging the different kinds of illative conversion in tabular form, we shall simplify them for reference. Taking the letter *p* to indicate *conversion by limitation* or *per accidens*; *s, simple conversion;* and *k, conversion by negation,* we shall have the following table:

ILLATIVE CONVERSION.

Original Propositions.	Methods of Converting.	Converted Propositions.	
(A) All X is Y.	*p.*	Some Y is X.	(I.)
(E) No X is Y.	*s.*	No Y is X.	(E.)
(I) Some X is Y.	*s.*	Some Y is X.	(I.)
(O) Some X is not Y.	*k.*	Some (not Y) is X.	(I.)

The above are the regular forms of conversion, but there are certain *Additional conversions* to be noticed. It must be remarked that the universal affirmative,

All X is Y = *All men are mortals,*

is sometimes converted in another manner; *i. e.,* by putting immediately before both subject and predicate the negative particle *not,* and then converting; thus,

All (not) Y is (not) X = *All (not) mortals are (not) men;*

i. e., All (who are not) mortals are not men; or, in common phrase, *None but Y can be X* = *none but mortals can be men.*

Again (E), which is converted simply, may be likewise converted *by limitation,* since, if having the universal form,

No A is B = *No men are trees,*

we can say

No B is A = *No trees are men,*

we can also say, what is less than this,

Some B is not A = *Some trees are not men.*

It may happen that for some purpose of logical technicality it will be better to use the *particular* when we have a right to use the *universal,* but from the existence of the *universal* we infer that of the *particular*, which is only a part of it.

There remains only one remark to be made upon the subject of conversion; it is that there are a few propositions which bear the form of *A* or *universal affirmative*, which are capable of simple conversion. The terms of such a proposition are said to be *convertible terms*, or the predicate and subject are either *exactly equivalent* or *exactly co-extensive;* for example, in the proposition *All common salt is chloride of sodium*, we have a right to assert that *all chloride of sodium is common salt.* From the proposition *All the good are saved*, we have a right to infer that *All (who are) saved are good.* Many *just definitions* come under this class. Besides such propositions as these, there are many mathematical propositions which seem to be single propositions with convertible terms, when in reality they contain two distinct propositions, each of which requires distinct proof. Thus, *All equilateral triangles are equi-angular.* The apparent converse that *All equi-angular triangulars are equilateral*, is indeed true, but this is not inferred from the original proposition; it is proved separately by geometricians; so that, instead of being the converse of the proposition stated, it is, in reality, a distinct proposition.

The processes of conversion have been applied above only to the forms of simple categorical propositions; they may likewise be applied, however, to compound propositions, and, when we come to consider these, we shall show how they may be converted; but it may be here observed, that as all compound propositions may be readily reduced to the simple categorical form, having shown how to convert these, we have in reality shown how to convert them all.

The next process of importance in considering propositions is the manner and character of their *opposition* to each other, and this, like the process of conversion, becomes of special value when we are joining propositions together to frame arguments.

(31.) Of Opposition.

Two propositions are said to be opposed to each other when, *having the same subject and predicate, the one denies either entirely or in part what the other affirms, or affirms either entirely or in part what the other denies;* as, for instance, the proposition

(A) *All men are mortal* is opposed by both $\begin{cases} \textit{No men are mortal.} & \text{(E.)} \\ \textit{Some men are not mortal.} & \text{(O.)} \end{cases}$

and (E) *No angels are men* is opposed by both $\begin{cases} \textit{All angels are men.} & \text{(A.)} \\ \textit{Some angels are men.} & \text{(I.)} \end{cases}$

Again, two propositions are said to be opposed when, *having the same subject and predicate, the one affirms in whole what the other affirms in part, or denies in whole what the other denies in part;* thus,

(A) All men are mortal.   (*Opp.*)   Some men are mortal.   (I.)
(E) No men are trees.   (*Opp.*)   Some men are not trees.   (O.)

Or, the rule may be more concisely stated thus: two propositions are said to be opposed to each other when, *having the same subject and predicate, they differ in quantity or in quality, or in both.*

It will appear, then, that the opposition in propositions is both in *quantity* and in *quality*, and as there are four forms of categorical propositions, and any two may be thus opposed, we shall have four kinds of opposition, which will best be illustrated by the following figure:

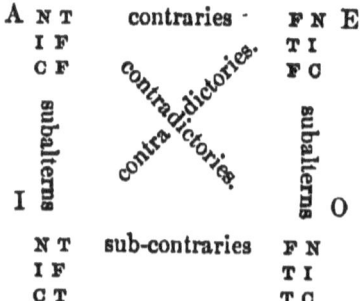

In which the two universal propositions A and E are called *contraries* and differ only in *quality*, being respectively *affirm-*

ative and *negative;* the two particulars I and O are called *sub-contraries,* differing likewise in *quality* only; the two *affirmatives* and the *two negatives* are called respectively *sub-alterns,* differing in quantity only; the *universal affirmative* and *particular negative,* and the *universal negative* and *particular affirmative,* are respectively called *contradictories,* and differ both in quantity and quality.

If we desire, as in applying Logic we may do, to determine the relative truth and falsity of these respective propositions, we must look for a moment at the *matter* which they may contain.

(32.) **Of the Matter of Propositions.**

The *matter* of a proposition is the *nature of the connection between the terms* of the proposition, or, in ordinary language, the *exact meaning of the proposition.*

By considering the nature of this connection between the terms, we shall see that it can be of only three kinds: *necessary,* which is expressed by an *affirmative* proposition; *impossible,* expressed by a *negative* proposition, and *contingent,* which is expressed by a *particular* proposition.

To illustrate: if we have given to us the two terms, *men* and *mortal,* and are told to connect them by a copula, we ask ourselves, what is the nature of the connection between these two? The answer is, it is *necessary,* and we express that *necessity* by using an affirmative copula, and prefixing the sign *All:*

All men are mortal.

Again, if we have given to us the two terms *men* and *trees,* to perform an analogous operation, we shall assert the nature of the connection between them to be *impossible,* and express that impossibility by the use of the prefix *No*—

No men are trees.

If, again, we have the terms *men* and *handsome,* we assert the nature of the connection to be *contingent,* as *some* men *are* and *some* are *not handsome;* and thus to express contingent matter we write the proposition with the prefix *some:*

> Some men are handsome.
> Some men are not handsome.

If, now, we examine the matter of these propositions we shall see that,

In necessary matter, all *affirmatives* are *true* and *negatives* false.

Necessary Matter.

True.	*False.*
(A) All men are mortal.	(E) No men are mortal.
(I) Some men are mortal.	(O) Some men are not mortal.

In impossible matter all *negatives* are *true* and *affirmatives* false.

Impossible Matter.

True.	*False.*
(E) No men are trees.	(A) All men are trees.
(O) Some men are not trees.	(I) Some men are trees.

In contingent matter all *particulars* are *true* and *universals* false.

Contingent Matter.

True.	*False.*
(I) Some men are handsome.	(A) All men are handsome.
(O) Some men are not handsome.	(E) No men are handsome.

From this examination we perceive that if one *contrary* is *true* the other must be *false,* but if *one* is *false* the other *may be false* also; if one *sub-contrary* is *false* the other must be *true,* but if one is *true* the other *may be true* also. But in the case of *contradictories,* if one is either *true* or *false,* the other must be just the *opposite, i. e., false* or *true.*

It remains to consider the *subalterns,* which differ in *quantity.* If the universal (A or E) be true, the particular (I or O) will be true also; as

(A) All men are mortal,	(E) No men are trees,
implies	implies
(I) Some men are mortal.	(O) Some men are not trees.

OF COMPOUND PROPOSITIONS. 81

If the particular I or O be true, the universal A or E is not *necessarily* true.

(I) *Some islands are fertile* does not permit us to infer (A) *All islands are fertile.*

(O) *Some islands are not fertile* does not permit us to imply (E) *No islands are fertile.*

But if the *particular* be *false*, the universal must of necessity be false also. Thus, the false particular *Some men are trees* would give us also *All men are trees* as a false universal.

By summing up these inferences we may state the following rules, which must be kept in the memory as we approach the subject of Reduction.

I. *Contraries* may both be *false*, but never both be *true*.

II. *Sub-contraries* may both be *true*, but never both *false*.

III. Of *Contradictories*, if one be *false* the other must be *true*, and *vice versa*.

IV. In *Subalterns* we reason from the *affirmation* only of the *universal* to the affirmation of the particular; but from the *denial* of the *particular* to the *denial* of the *universal*.

The letters N I C at the corners of the figure indicate *necessary, impossible* and *contingent* matter; T means *true*, and F false.

The passage from one proposition to another in *conversion* and *opposition* is called by some writers *immediate inference.*

With the remark that opposition may be also illustrated in compound propositions, or those not directly in the simple categorical form, or that such propositions may be reduced to this simple form by an easy process still to be explained, we pass to the subject of compound propositions.

(33.) Of Compound Propositions.

A *compound proposition* consists of two or more simple propositions, united together either by a simple copulate, expressed or understood, or by a conjunction denoting an hypothesis.

F

Compound propositions are consequently divided into two classes, *categorical* and *hypothetical*.

Compound categorical propositions are of two kinds, *copulative* and *discretive*.

A *copulative* proposition consists of two or more subjects united with the same predicate, or with two or more predicates, by the use of the *copulative* conjunction; as,

>Men, horses and birds are animals.

A *discretive* proposition consists of two simple propositions, which are contrasted on account of an apparent inconsistency; as,

>Fox, though dissolute, was a patriot.

In this a third proposition is implied, viz., the general incongruity of patriotism with dissoluteness.

Many compound propositions are *tacit* or *implied*, and thus have the form of *simple propositions*.

A *hypothetical* proposition consists of two or more simple propositions united by a conjunction which expresses hypothesis. This conjunction is usually placed at the beginning of the proposition.

Hypotheticals are divided into *conditional, disjunctive* and *causal*, and take these names from the conjunctions which express the condition of the hypothesis.

A *conditional* proposition expresses the condition by the conjunction *if*; as,

>If A is B, C is D = If John return, Harry will go.

A disjunctive proposition is formed with the conjunctions *either* and *or*; as,

>Either A is B, or C is D = Either John is wrong, or James is ill.

A causal proposition unites its parts by the conjunction *because*; as,

>A is B because C is D.
>John is well because he is prudent.

OF COMPOUND PROPOSITIONS.

It is evident, in the case of categorical propositions, that they may be at once resolved into the simple propositions of which they are composed: thus we may divide the *copulative* proposition given into three distinct propositions, viz.,

> Men are animals,
> Horses are animals,
> Birds are animals.

and the *discretive* may be divided into two, thus:

> Fox was dissolute,
> Fox was a patriot.

Unlike the compound *categorical* propositions, the *hypotheticals* contain within themselves the germ of an argument, and only require *that the hypothesis shall be established*, or *fail* of establishment, to arrive at a conclusion. Thus, having the proposition,

> If A is B, C is D,

we need only know whether A *is* B, in order to state the argument and arrive at the conclusion that C *is* D.

Conditional propositions, however, may be, in every case, reduced to a categorical form, by regarding them as *universal affirmative* categorical propositions, of which the *antecedent* is *the subject*, and the *consequent the predicate*. We then rid ourselves of the *condition*, by the use of the words "the case of;" thus, instead of the form, If A is B, C is D, we shall have

> (*The case of*) A being B, *is* (*the case of*) C being D,

which is purely categorical in form.

Disjunctive propositions may be reduced to *conditionals*; thus:

> Either A is B, or C is D, is equivalent to if A is not B, C is D,

or we may place it at once in a categorical form without this double process, by reading it thus:

> *The two possible cases in this matter are tha.. A is B, and that C is D.*

It is more usual to reduce the disjunctive, however, to a conditional form, into which it very naturally falls.

The *causal proposition,*

Because A is B, C is D,

becomes either at once categorical, when we establish the truth of *because,* and thus we have

A is B, therefore C is D,

as an enthymeme, to which, having the subject-matter, we might supply the wanting premiss; or the *causal proposition* becomes simply *conditional,* if the *cause*—expressed by the first proposition *A is B*—be doubtful, and then we read,

If A is B, C is D,

which must be treated like the *conditional* above.

As it seems, then, that all these are reducible to the *conditional* form, we need only show how the process of conversion is applied to conditionals, in order virtually to apply it to them all. From what has been said, it will appear that conditionals are converted *by negation* only; thus, to convert the proposition,

If John has the smallpox he is sick,

we may read—

If John is *not sick* he has *not* the smallpox;

or, the conversion rests upon the fact that the *denial of the consequent* leads to the *denial of the antecedent.*

We cannot convert without this negation, for we could not reason from the *affirmation* of the *consequent* to the *affirmation of the antecedent;* thus,

If John is sick he has the smallpox,

since that *consequent* (*sickness*) may have sprung from some *other antecedent* than *the smallpox.*

(34.) The New Analytic.

And here it becomes necessary, before closing the subject of *propositions,* to refer briefly to the effort of certain late writers to *quantify* the *predicate;* that is, to place prefixes before it similar to those placed before the *subjects* of propositions to determine at a glance its distribution or non-distribution, and to form thus a new set or class of categorical propositions. Thus, instead of the form *all men are animals*, they would write *all men are some animals*, and claim thereby not only a greater precision in the logical statement, but in some instances the establishment of a distinct proposition; as, for example,

All A is (all) B.

It may be admitted that sometimes a new idea is suggested by such a quantification of the predicate, but it is only *suggested*, not contained in the proposition thus rendered. Thus, if we say,

All men are sinners,

we mean by our rule, *some sinners;* now the question as to the comprehension of this word *sinners* may arise, when we place such a prefix; whether *angels* and *devils* may or may not be included in it; and whether the *ill-conduct of brutes* is excluded from it. Whereas, if we could write,

All men are (all) sinners,

we should exclude at once all other beings from the category. Hence, the quantification of the predicate, which in the old system is implied, does, when expressed, suggest new thoughts or judgments, but those new judgments rest upon their own basis, and have really nothing to do with the original proposition. There seems really, therefore, nothing gained in the extension of the proposition by this attempt to quantify the predicate, but rather a confusion of judgment and a complication of logical forms.

It is not intended to give, in detail, the applications of the "new analytic," nor to deny that results, totally out of the province of Logic, are attained by it. It is evident that if we quantify the predicate, in categorical propositions, we shall have four additional forms, viz.:

	Established Forms.	*New Forms.*	
A.	All A is B.	All A is all B.	X.
E.	No A is B.	No A is some B.	Y.
I.	Some A is B.	Some A is all B.	U.
O.	Some A is not B.	Some A is not some B.	Z.

Now of these new forms we have already considered X, as in the case,

All equilateral triangles are (all) equi-angular,

and in the cases of exact definitions, as

All common salt is (all) chloride of sodium.

In the first we have seen that there are *two distinct propositions*, and in the second that there are but *two names for the same object*.

As for Y, U and Z, they are so clearly contained in the old forms that they need but little elucidation.

    U.   *Some trees are all oaks,*

when converted gives us

        All oaks are trees,   or A.
Y.   No heroes are some men.
Conv. Some men are not heroes.   O.
Z.   Some quadrupeds are not some horses.

By which we determine that the quadrupeds referred to may belong to other species, or may be included in the species *horse*, apart from the *some horses* mentioned.

It was attempted, in the new analytic, to simplify the subject of conversion, but, it seems, with inadequate results.

And here we leave the subject of quantifying the predicate, so far as it relates to propositions alone. If carried out in the syllogism, it would much enlarge the domain of Figure, and give much fruitless labor to the logician.

CHAPTER VII.

(35.) Of Arguments.

AN *argument* is an act of reasoning or *ratiocination*. It consists of two parts: that to be proven, and that by which it is proven.

The part to be proven is embodied in the *conclusion*, and that by which it is proven is embodied in the *premisses*. When these are inverted from the usual logical order, so that the *conclusion* is stated first, it is called the *question;* and the premisses which are joined to it by the word *because*, are then called *the reason;* thus,

(Question) *Why are all Americans mortal?*
    or *All Americans are mortal,*
Because They are men.

But in logical form and order the premisses are stated first, and the conclusion is connected with them by the illative conjunction *therefore;* thus,

Premisses { All men are mortal,
               All Americans are men.
Therefore   All Americans are mortal.

These two forms must be distinguished from what is expressed by the words *inference* and *proof*, which have not to do with the order of the parts in an argument, but with the special design of the person who uses the argument; *i. e.*, whether from known facts or *premisses*, he seeks to establish a *conclusion;* or has *adopted a conclusion*, and is simply seeking for premisses by which to substantiate it.

Logic teaches us to draw from known proofs only a *just inference*, or to maintain a given inference only by *just proofs*.

We may more clearly illustrate by observing how, in the various professions, these different methods are used; thus, a naturalist gets together many observations and makes many experiments, forming a strong store of *proofs*, before he may justly *infer a conclusion*, while an advocate at law *assumes* the innocence of his client or the guilt of the prisoner, *as a foregone conclusion*, and then uses every means for *obtaining proofs* and thus *establishing premisses* by which to substantiate his conclusion.

It has been observed that the logical form of an argument is a *syllogism*, which consists of three propositions; *i. e., two premisses and a conclusion*.

After fully explaining the *syllogism*, we shall consider all forms of *irregular* and *abridged arguments*, and show, as has been asserted, that they may all be reduced to this simple form, so that the logical tests may be at once applied to them.

(36.) Of the Syllogism.

In the analysis of Logic, the dictum of Aristotle was distinctly laid down and illustrated. Its form was:

No. 1.	*No.* 2.
All A is B.	No A is B.
All or some C is A.	All or some C is A.
All or some C is B.	No C is B, or some C is not B.

The principle of the dictum is, that *whatever* (B) we predicate (*in the major premiss*) of the whole class (All A); under which class we assert (*in the minor premiss*) certain individuals (All or some C) to be ranged; we may also predicate (*in the conclusion*) of those individuals.

Thus, B is predicated of (All A), C is an individual of the class A, therefore we have a right to predicate B of C.

But, as few arguments, in the ordinary uses of language, are placed in this exact form (although all valid arguments may be), there have been laid down two logical axioms and several important rules for determining the validity of syllogisms without the labor of bringing them to this form.

It must be constantly remembered that it is a condition of every syllogism that it contains *three* and only *three* terms: the *major* term, the *minor* term, and the *middle term.* The first two of these terms must not be confounded with the *premisses* which bear the same name, and which are *propositions.* Thus in the example:

	mid.	maj.	mid.	maj.
Maj. prem.	A is B	=	All men	are mortal.
	min.	mid.	minor.	mid.
Min. prem.	C is A	=	All Americans	are men.
	min.	maj.	minor.	major.
Concl.	C is B	=	All Americans	are mortal.

B is the *major* term, and it is in the major premiss; C is the *minor* term, and it is found in the minor premiss; A is the *middle* term, because it is the *medium* of comparison between the other two. In the *major premiss,* the *middle* term is compared with the *major;* in the *minor premiss* it is compared with the *minor,* and *in the conclusion,* the *minor* and *major* terms, having been thus found to agree with the same *middle* term, are asserted to agree with each other.

The *minor* term is *always* the *subject* of the *conclusion,* and the *major* term the *predicate.*

This simple process of comparison leads us to the statement of those axioms which determine the conditions of agreement and disagreement between the major and minor terms, and to note some important consequences following from them.

(37.) Logical Axioms.

1st. If two terms agree with one and the same third term, they will agree with each other.

2d. If of two terms, the one agree and the other disagree with one and the same third term, they will disagree with each other.

Rules.

I. From the first of these axioms we observe that if both premisses of a syllogism are *affirmative,* thus expressing the agreement of the *major* and *minor* terms with the *middle*, the conclusion must likewise be *affirmative*, or express the agreement between these two terms; thus, B being the major term, C the minor, and A the middle, we have

>A is (or agrees with) B,
>C is (or agrees with) A,

and we must consequently state the conclusion

>C is (or agrees with) B.

II. Again, from the second axiom, we see that if one of the premisses (*as the major*) be *affirmative*, and thus express the agreement between the *major* term and the *middle*, and the other be negative and thus express a disagreement between the *minor* term and the *middle*, we must have a *negative* conclusion to express the disagreement between the *major* and the *minor*, which we have thus shown, the one to agree and the other to disagree in the premisses with one and the same third (*the middle*).

Thus, if A is not (*or disagrees with*) B,

And if C is (*or agrees with*) A,

we must have, C is not (*or disagrees with*) B.

III. It is further evident that *if both premisses be negative, we can draw no conclusion;* because in these premisses the *middle* term, simply disagreeing with both the *major* and *minor* terms, is no longer a medium of comparison between them. For example, state the premisses,

>No A is B = No men are trees,
>No C is A = No horses are men;—

we have established no relation whatever between C and B, or between *horses* and *trees*, so that, although we might *truthfully* write

>*No horses are trees,*

it would be an accidental statement, and not spring from the premisses stated.

In the conclusion is stated the *relation between the major and minor term,* which was established in the premisses by the medium of the middle term. The *minor* term is the true *subject* of the conclusion, and the *major* term the true *predicate.* Sometimes in an inverted or elliptical conclusion these terms may appear transposed, but when properly written out they will take the places indicated.

The *middle term,* which occurs twice in the premisses, is the medium of comparison between the two other terms, and is generally the *name of a class,* of which in *one* premiss something is predicated, or to which some quality is attributed, as

   1. *Man* is a rational animal,

in which *man* is the name of a class, and *rationality* a predicate or attribute: under which in the other premiss we range an individual or individuals belonging to the class, as

   2. *John* is a *man,*

and by means of which we have a right to predicate or attribute this same thing rationality to the individual; thus,

   3. *John* is a *rational animal.*

IV. *Ambiguous middle.*

It is scarcely necessary to state that the middle term must be *univocal, i. e.,* must have the same meaning in both premisses. If it be *ambiguous,* or possess one meaning in the *major* premiss and a different one in the *minor,* we shall violate the first principle in the construction of a syllogism, and have *four* terms instead of the *three, and only three,* required. Most languages have many such ambiguous words, and the English particularly is full of them: thus

   1. A *bank* is a financial institution.
   2. The margin of a stream is a *bank.*
   3. The margin of a stream is a financial institution.

Many such glaring examples will occur at once to the student; but it must be remembered that the sophist who would construct his artful fallacies to deceive, does not employ such manifestly ambiguous words, but those whose double meanings are much more nearly the same.

Thus, in their philosophic meanings, the words *church* and *faith* have given rise to sharp controversy and violent partisanships. As ambiguous terms play a very prominent part in the subject of Fallacies, we shall recur to them under that head.

When the argument is written out in symbols, the ambiguity either disappears entirely, that is, when we represent the term in both premisses by the same letter, thus,

$$A \text{ is } B,$$
$$C \text{ is } A,$$
$$C \text{ is } B,$$

or it becomes at once manifest, when we represent the term in the major premiss by one symbol, as A, and that in the minor, having a different meaning, by another, as D, thus,

$$A \text{ is } B,$$
$$C \text{ is } D,$$

in which premisses there are four terms, and the error distinctly appears.

V. *Undistributed middle.*

The middle term must be *distributed; i. e.*, taken in its whole comprehension, *at least in one of the premisses*, for it will otherwise occur that we may compare the *major* term with *one part of the middle*, and the minor with *another part*, and thus it would fail to be a just medium of comparison. It might happen, by chance, that these two parts should be the *same*, but it would be only by *chance;* in the general case they would be different parts, and if we choose to regard *each part* as a *distinct term*, we should again run into the error of having four terms instead of *three;* thus,

> Some quadrupeds are cows,
> Some quadrupeds are sheep,
> Therefore Some sheep are cows.
>
> White is a color,
> Black is a color,
> Therefore Black is white.

But if one of the extremes be compared with the *whole* of the middle term, and the other be compared *only with a part*, which *part* is necessarily contained in the *whole*, they may then be compared with each other.

VI. *Illicit process.*

Again, in order to *distribute* either the *major* or *minor* term in the conclusion, it must have been previously distributed in the premiss in which it occurs: because, we only have a right to compare *that part* of the term with the other, in the conclusion, which we have already compared with the *middle* in the premiss; thus,

All men are animals,
No dogs are men,
Therefore No dogs are animals.

The technical name for this logical fallacy is the *illicit process.* In the example, the *major* term, *animals*, which is not distributed in the premiss (as it is the predicate of an affirmative proposition) is distributed in the conclusion (as the predicate of a negative proposition); this is called an *illicit process* of the *major* term; if it be the *minor* term thus treated, it is called an *illicit process* of the *minor* term.

The following is an example of illicit process of the *minor*.

1. All men are rational beings,
2. All men are animals,
3. All animals are rational beings.

In this example the minor term *animals*, which is undistributed in the minor premiss—as the predicate of an affirmative proposition—is distributed in the conclusion, being there the subject of a universal.

Let it be remembered that this is called an illicit process of the major or minor *term*, not of the major or minor *premiss*.

VII. If both premisses in a syllogism be *particular* propositions, we can draw no conclusion; thus,

1. Some men are wise,
2. Some men are foolish,

leads us to no conclusion. Nor are we benefited if we make one of the premisses *particular negative;* thus,

1. Some men are wise,
2. Some men are not brave,

we are as before without any medium of comparison.

The fact is as stated; the causes are various, and will be fully explained in the chapter on *Figure.*

It is sufficient, now, for the student to know that the cause is in every case either an *undistributed middle* or an *illicit process* of one of the other terms.

By the foregoing axioms and rules, we extend the range of syllogistic forms, and are able to see the validity or invalidity of an argument without reducing it to the invariable formula of Aristotle's dictum. We proceed now to show how many of these forms there may be, and the relation they sustain to the *dictum* itself; and this brings us to the subject of *Figure* and *Moods.*

CHAPTER VIII.

OF FIGURE AND MOODS.

(38.) Figure.

Figure is the technical name employed to designate the classification of syllogisms according to the *position of the middle term with reference to the two extremes* in the premisses. Now, it is evident that the middle term can have only four variations of position, and hence we say there are *four figures*.

1st. The middle term may be the *subject* of the *major* premiss, and the *predicate* of the *minor*, and this designates the 1*st figure*.

2d. It may be the *predicate* of *both premisses*, and thus the 2*d figure* is designated.

3d. In the 3*d figure* it is the *subject* of *both premisses;* and

4th. In the 4*th figure* (which is the reverse of the 1st) it is the *predicate* of the *major premiss* and the *subject* of the *minor*.

If we designate the *major* term by P (as it is always the *predicate* of the *conclusion*), the *minor* term by S (being the *subject* of the *conclusion*), and the *middle* term by M, and merely state these various positions of the middle term, without considering or denoting the quantity or quality of the propositions in the syllogism, we shall have the abstract syllogism,

I.	II.	III.	IV.
M is P.	P is M.	M is P.	P is M.
S is M.	S is M.	M is S.	M is S.
S is P.	S is P.	S is P.	S is P.

These are called the *four figures;* and to the syllogisms which occur in them the axioms and rules already laid down directly apply.

If now we proceed to examine these figures in order, we shall find that the first figure is but the symbolical representation of Aristotle's *dictum*, the simplest form of the syllogism. There will be four variations of it, viz.:

1.	2.	3.	4.
All M is P.	All M is P.	No M is P.	No M is P.
All S is M.	Some S is M.	All S is M.	Some S is M.
All S is P.	Some S is P.	No S is P.	Some S is not P.

We have simply supplied the quantity and quality required.

Since, in the major premiss, then, of Aristotle's dictum, we *assert or deny the predicate* of the *whole class which is the subject* (All M), it is evident that in the *first figure*, the *major premiss is always universal.* If, then, with this relative position of the middle term, *i. e., in the first figure*, we find a syllogism the *major premiss* of which is *particular*, we may at once declare it to be *invalid*.

Again, since the province of the *minor* premiss in *the dictum* is always to *assert* that certain individuals belong to the given class (and in no case to deny it), it appears that in the first figure the minor premiss must always be *affirmative*, so that if we find a syllogism in this figure with a *negative* minor premiss, we may at once declare it *invalid*.

Thus, in stating the four forms of the dictum, we have stated the only four forms which the first figure can cover.

But the other figures, which are not directly in the form which the dictum assumes, instead of being explained by it, are to be considered in the light of the axioms and rules for determining the validity of syllogisms when the dictum does not directly apply. By examining the *second figure*,

P is M,
S is M,
S is P,

we shall find that there are several forms which it will assume when we supply the quantity and quality to the propositions.

We observe at once that the conclusion must in every case be *negative*, because—

1st. The *middle* term is the *predicate of both premisses*.

2d. The *middle term must be distributed at least once* in the syllogism.

3d. In order that the *predicate* of a proposition shall be *distributed*, the proposition must be *negative*.

4th. This will give us one negative *premiss*, and by the second axiom, if we have a negative premiss, the *conclusion* must be *negative (universal or particular)*.

Third Figure.
M is P,
M is S,
S is P.

By the supplying of quantity and quality, this figure assumes a greater variety of forms than any other.

By considering the position of the terms here, it will appear that we can only draw *particular* conclusions. For if both premisses be affirmative, and we draw a *universal* conclusion, or *All* S is P, then S (the minor term), which was *undistributed* in the *minor premiss* (being the predicate of an affirmative proposition), will be *distributed* in the *conclusion*, as the subject of a universal; or we shall have an *illicit process of the minor*.

If the major premiss be negative, and we draw a *universal conclusion*, it is easily shown that the same error—an *illicit process of the minor*—obtains; and if the minor premiss be negative, we shall have an *illicit process of the major*.

Fourth Figure.
P is M,
M is S,
S is P.

The fourth figure, which was not proposed by Aristotle with the other three, and only recently adopted by logicians, is an inversion of the first, and an unnatural and unnecessary form of the syllogism. By a similar examination of all the terms, we shall find that we may draw, as conclusions, in this figure all the categorical propositions except *A*, which, as has been shown, can only be drawn in the first figure. It is the prerogative of Aristotle's *dictum* alone, to draw from certain premisses a *universal affirmative* conclusion.

The various forms of the syllogism due to the different quantity and quality of the propositions composing them are arranged, in the different figures, in what are called *moods*, or a concise manner of expressing a syllogism by symbols.

(39.) Of Mood.

If, having any syllogisms, as the following—

1. $\begin{cases} \text{All A is B,} & (A) \\ \text{All C is A,} & (A) \\ \text{All C is B,} & (A) \end{cases}$
2. $\begin{cases} \text{No A is B,} & (E) \\ \text{Some C is A,} & (I) \\ \text{Some C is not B,} & (O) \end{cases}$

we write together the symbols characterizing each proposition which composes them, we are said to determine the mood of the syllogism; thus, the symbol of the major premiss in the first syllogism is

A, or universal affirmative;

that of the minor,

A, or universal affirmative;

and that of the conclusion likewise

A, or universal affirmative.

Hence we say that *A A A* is the mood of the syllogism.

In the second syllogism, we shall find by a similar process that the mood is *E I O*.

Now, it is evident that the number of moods we can have will depend upon, 1st, the number of propositions in the syllogism, viz., *three;* and 2d, upon the number of categorical propositions which we can enumerate, viz., *four*, A, E, I, O;

it becomes then a simple algebraic arrangement of four letters, A, E, I, O, in three columns *in every possible combination*. The number of these possible combinations will be sixty-four. For each of the propositions A, E, I and O may have a major premiss; and each of these may have each in turn as a minor premiss; thus,

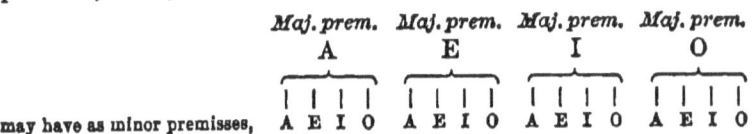

Again, each of these sets (sixteen in all) may have four different conclusions, *i. e.*, each of the categoricals as a conclusion. Taking the first set, for example, and supposing the operation performed for the rest:

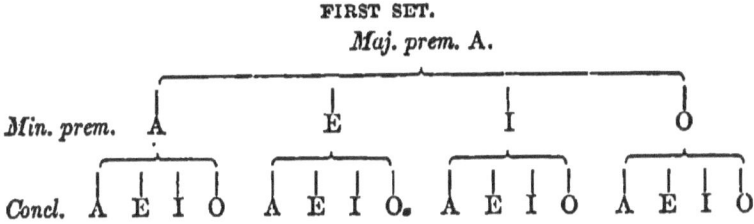

This same process may be performed for E, I and O. There will evidently be *sixty-four* moods, of which, however, it is at once evident that very many will violate the axioms and rules already laid down, and must be for this reason discarded.

Thus, all the combinations of *affirmative* premisses having negative conclusions, as A A E, A I O, etc., etc., must be thrown aside, because they violate the first axiom.

All the sets of *negative* premisses, with whatever conclusions, are useless, as E E, O O, E O, O E, etc.

All the sets of particular premisses, with whatever conclusions, must be neglected, such as I I, O O, O I, I O, etc.

If all these eliminations be performed—and, simple as they are, the student is advised to go carefully through them *once* for himself—we shall find twenty-eight moods excluded on account of negative and particular premisses: eighteen by the condition that the conclusion follows the inferior part, and we shall see that one—I E O—is rejected for an illicit process of the *major* term, in every figure, and finally that of the *sixty-four arrangements* which we call *moods,* only *eleven* represent *valid arguments,* or

FOUR AFFIRMATIVES AND SEVEN NEGATIVES.

A A A	E A E
A I I	A E E
A A I	E A O
I A I	A O O
	O A O
	E I O
	A E O

If now we apply these moods to each figure, in detail, it would seem, since there are *four* figures, that we should have $4 \times 11 = 44$ moods in all the figures; but in this application we find that many moods which are valid in one figure are not in others; as, for example, the mood I A I, which is allowable in the third figure, would be in the first figure a case of *undistributed middle,* and would further violate the principle of Aristotle's dictum, which requires that the *major* premiss should be a *universal* proposition. A E E is a valid mood in the *second figure,* while, in the *first,* it would have an illicit process of the major term, and would further violate that principle of the dictum which requires the *minor* premiss to be always *affirmative.*

By applying these eleven moods to the four figures, we find that there would be *six* in *each figure,* or *twenty-four in all;* but even of these, *five* are omitted as useless; for example, the mood A A I, in the first figure, because it is implied and contained in the mood A A A. Since, if the universal conclusion A be true, the particular I is necessarily true. By an application of each of these moods to every figure, we shall have left, finally, *nineteen* moods in all; or, FOUR *in the first figure,* FOUR *in the second,* SIX *in the third,* and FIVE *in the fourth.*

The moods of the first figure are called *perfect* moods; those in the other figures, *imperfect* moods.

As it has been asserted that all arguments may be put in the form of Aristotle's *dictum,* that is, that all the *imperfect* moods may be made *perfect,* we proceed to fulfill this assertion, by the process of *reduction, i. e.,* the reducing of moods in the 2d, 3d, and 4th figures to the 1st figure, which is the form of the dictum.

In order to facilitate this process, as well as to retain easily in the memory the different moods and their value, the following verses, Latin in sound and scansion, but without intrinsic meaning in the words, have been formed:

FIG. I.—BArbArA, CElArEnt, DArII, FErIO, *dato primæ.*
FIG. II.—CEsARE, CAmEstrEs, FEstInO, FAkOrO, *secundæ.*
FIG. III.— { *Tertia* DArAptI, DIsAmIs, DAtIsI, FElAptOn, DOkAmO, FErIsO, *habet; quarta insuper addit.*
FIG. IV.—BrAmAntIP, CAmEnEs, DImArIs, FEsApO, FrEsIsOn.

There are variations in these lines, made by various writers; we have adopted the above as the form which will indicate to us in the simplest manner the processes of *Reduction.*

Before explaining these lines, which the student must memorize in order to make them useful, that he may have the moods, and their places in the figures, at his tongue's end, it will be observed that there are a few words used in these verses which are of no use *except to make out the hexameter lines;* of these are *dato primæ* in the first, *secundæ* in the second, *tertia habet* in the third, and *quarta insuper addit,* which states—*moreover the fourth adds,* etc. Leaving these out of the consideration, in the lines themselves the *vowels* in each word represent the *moods;* thus, *Barbara* is the mood *A A A; Cesare,* the mood *E A E,* etc., etc.

The following *consonants* indicate what changes are to be

made in the given *imperfect mood* to reduce it to a *perfect mood* of the *first figure:*—*s*, that the proposition indicated by the vowel immediately preceding it is to be *converted simply;* thus in *Camestres*, the first *s* indicates the simple conversion of the first *E*, or the minor premiss, and the last *s* the simple conversion of the second *E*, or the conclusion. In similar relations *p* and *k* stand respectively for *conversion by limitation* and *conversion by negation; m*, wherever it occurs, expresses that the premisses must be transposed; the other consonants have no meaning, and are only employed to frame the words. P, in the mood *Bramantip* of the fourth figure, denotes that the transposed premisses, indicated by *m*, will warrant a universal conclusion instead of a particular. The initial letters, B, C, D, F, of the words which contain the moods, are so arranged throughout the figures as to indicate the mood in the *first* figure to which any imperfect mood *will be reduced;* thus *Darapti* of the third figure will, when reduced, become *Darii* of the first, *Camestres* will become *Celarent*, etc.

It must be observed that this arrangement is only for the sake of convenience, as the process of reduction is invariable, and the mood *Darapti* would become, when reduced, the mood A I I of the *first* figure, whether it were called *Darii* or by some other name. Students are apt to be misled with reference to these initial letters, and to suppose that they will aid them in the process of reduction. It is on this account that they are cautioned that this is only a convenient and not an auxiliary arrangement. Before proceeding to explain the system of reduction, let us give an example of each mood, in all the figures, putting the logical frame-work to its legitimate use, and showing every form which the syllogism can assume. We shall make the examples very simple, leaving it to the student, with these before him, to frame longer and more complex ones for himself—a practical exèrcise which will be found very useful. The *middle* term is placed in *italics* in each example.

Examples.

FIGURE I.

Barbara.

A. Every *desire to gain by another's loss* is covetousness.
A. All gaming is a *desire to gain by another's loss.*
A. All gaming is covetousness.

Celarent.

E. No *one who is enslaved by his appetites* is free.
A. Every sensualist is *one who is enslaved by his appetites.*
E. No sensualist is free.

Darii.

A. All *pure patriots* deserve the rewards of their country.
I. Some warriors are *pure patriots.*
I. Some warriors deserve the rewards of their country.

Ferio.

E. *Nothing which impedes commerce* is beneficial to the revenue.
I. Some taxes *impede commerce* (or are *things which impede commerce*).
O. Some taxes are not beneficial to the revenue.

FIGURE II.

Cesare.

E. No vicious conduct is *praiseworthy.*
A. All truly heroic conduct is *praiseworthy.*
E. No truly heroic conduct is (or can be) vicious.

Camestres.

A. Every true philosopher *accounts virtue a good in itself.*
E. No advocate of pleasure *accounts virtue a good in itself.*
E. No advocate of pleasure is a true philosopher.

The true middle term here would be (*one who*) *accounts virtue a good in itself.*

Festino.

E. No righteous acts *will produce ultimate evil to the actor.*
I. Some kinds of association *will produce ultimate evil to the actor.*
O. Some kinds of association are not righteous acts.

Fakoro.

A. All true patriots are *friends to religion.*
O. Some great statesmen are not *friends to religion.*
O. Some great statesmen are not true patriots.

FIGURE III.

Darapti.

A. All *wits* are dreaded.
A. All *wits* are admired.
I. Some admired (persons) are dreaded.

Disamis.

I. Some *lawful things* are inexpedient.
A. All *lawful things* are what we have a right to do.
I. Some things which we have a right to do are inexpedient.

Datisi.

A. All *that wisdom dictates* is right.
I. Something *that wisdom dictates* is amusement.
I. Some amusement is right.

Felapton.

E. No *science* is capable of perfection.
A. All *science* is worthy of culture.
O. Something worthy of culture is not capable of perfection.

Dokamo.

O. Some *noble characters* are not philosophers.
A. All *noble characters* are worthy of admiration.
O. Some (who are) worthy of admiration are not philosophers.

Feriso.

E. No *false theories* exist in a perfect state of being.
I. Some *false theories* are harmless things.
O. Some harmless things do not exist in a perfect state of being.

FIGURE IV.
Bramantip.

A. All oaks are *trees*.
A. All *trees* are vegetables.
I. Some vegetables are oaks.

Camenes.

A. All men are *mortal*.
E. No *mortal* is a stone.
E. No stone is a man.

Dimaris.

I. Some taxes are *oppressive*.
A. *All (that is) oppressive* should be repealed.
I. Some things which should be repealed are taxes.

Fesapo.

E. No immoral acts are *proper amusements*.
A. All *proper amusements* are designed to give pleasure.
O. Some (things) designed to give pleasure are not immoral acts.

Fresison.

E. No acts of injustice are *proper means of self-advancement*.

I. Some *proper means of self-advancement* are unsuccessful.
O. Some unsuccessful (efforts) are not acts of injustice.

It will be observed that the conclusions in the fourth figure are indirectly stated, and that it would seem as if in tracing the *major* term back from its place as *predicate* of the conclusion, it is in reality predicated by means of the other terms of itself; thus, in the conclusion it is predicated of the *minor*, which in the *minor* premiss is predicated of the *middle*, which in the *major* premiss is predicated of the *major*. The fourth figure, therefore, is not often used, and is rather accidentally stumbled into than employed intentionally.

The exact accordancy of the *first* figure with the dictum of Aristotle has been already stated. Of the *second* figure, it may be remarked that it is commonly used to *disprove* something that has been maintained, or is likely to be believed, although not true. As an illustration, suppose it had been asserted that

> All great statesmen are true patriots.

Then our example just given of *Fakoro* would be a refutation of this, and the argument would naturally take that form.

Of the *third* figure, it will appear that it will be useful where we have *singular* terms, which can only be *subjects* of propositions—*i. e.*, never *predicates*—and also where our purpose is to offer and sustain an objection to our opponent's premiss, which is *particular* when the argument requires it to be *universal*.

There are very many inverted and curious forms of *arguments* growing out of the elliptical and inverted forms of *propositions*, which we have already considered. Two common examples of these are added by way of illustration.

1.
> None but whites are civilized.
> The Hindoos are not whites.
> The Hindoos are not civilized.

The phrase *none but whites* may be rendered, *other than whites;* and this being the true middle term, we shall have—

>No *other than whites* are civilized.
>All Hindoos are *other than whites.*
>No Hindoos are civilized.

Which is evidently a syllogism in *Celarent* of the *first* figure.

>2.
>No one is rich who has not enough.
>No miser has enough.
>No miser is rich.

The major and minor premisses must be put in the form of categorical propositions, and we shall have—

>No *one who has not enough* is rich.
>Every miser is *one who has not enough.*
>No miser is rich.

Which is likewise in the mood *Celarent.* In both these examples the minor premiss, which appears to be a *negative* proposition, is in reality *affirmative.*

(40.) Of Reduction.

If we have any *imperfect* mood—*i. e.*, a mood in the second, third, or fourth figure—and we desire to prove the same conclusion in the *first* figure, so that the dictum of Aristotle may immediately be applied to it, the process by which this is done is called *Reduction.*

Reduction is of two kinds, *direct* and *indirect.* Direct reduction consists in proving in a *perfect* mood either the same conclusion, or one which, being illatively converted, will give us the same conclusion which we had in the *imperfect* mood. *Indirect* reduction consists in proving, *not* that the *original conclusion is true*, but that its *contradictory is false*, from which—by the scheme of opposition—we know that the original conclusion must be true.

Of direct reduction.

It has been shown that we have a right to *convert* any of

the propositions of the syllogism illatively; and it is also evident that we may *transpose the premisses* without affecting the truth of the propositions or the validity of the argument. If, then, we apply the processes indicated by the letters in the mnemonic lines, we shall see that they will give us the forms of *direct reduction*.

Taking for example *Cesare*, the mood $E\,A\,E$ in the second figure; to write it out we remember in the first place that the position of the middle term in the *second* figure is *predicate* of both premisses, and we observe that the major premiss is E, universal negative, the minor premiss A, universal affirmative, and the conclusion E, universal negative; we have then X, being the *major*, Z the *minor* and Y the *middle* term—

Cesare. Fig. II.

E. No X is Y = No men are trees.
A. All Z is Y = All oaks are trees.
E. No Z is X = No oaks are men.

The only consonant in the word CEsArE which indicates a process of reduction is *s*, which tells us that the major premiss, expressed by the first E, is to be simply converted; performing this operation we shall have—

Celarent. Fig. I.

E. No Y is X = No trees are men.
A. All Z is Y = All oaks are trees.
E. No Z is X = No oaks are men.

This syllogism is in the first figure, since the middle term *Y* or *trees* has become the *subject* of the *major* and the *predicate* of the *minor* premiss; again,

Fakoro. Fig. II.

A. All X is Y = All good men are virtuous.
O. Some Z is not Y = Some warriors are not virtuous.
O. Some Z is not X = Some warriors are not good men.

The *k* expresses that the major premiss (A) is to be converted by *negation;* performing this operation (there is no other indicated), we shall have—

OF REDUCTION.

Ferio. Fig. I.

E. All (not Y) is not X = All (not virtuous) are not good men.
I. Some Z is (not Y) = Some warriors are (not virtuous).
O. Some Z is not X = Some warriors are not good men.

This process, in effect, changes our middle term from Y or *virtuous* to (*not Y*) or (*not virtuous*), while we have the same conclusion as before in the mood *Ferio* of the first figure.

The reduction of the other moods of the *second figure* will be analogous to those already performed, and the student will find no difficulty in reducing them for himself. Passing, then, to the *third figure*, and remembering that in this figure the middle term is the *subject* of *both premisses*, let us reduce the mood

Disamis. Fig. III.

I. Some Y is X = Some men are heroes.
A. All Y is Z = All men are mortal.
I. Some Z is X = Some mortals are heroes.

The two letters which indicate changes in the process of reducing this mood are *s* (twice employed) and *m*: *s* indicates the simple conversion of the major premiss and the conclusion, and *m* the transposition of the premisses; performing these operations, we have

Darii. Fig. I.

A. All Y is Z = All men are mortal.
I. Some X is Y = Some heroes are men.
I. Some X is Z = Some heroes are mortal.

Which conclusion is the simple converse of the original conclusion, as was indicated by the final *s*.

Fesapo. Fig. IV.

A. No X is Y = No quadrupeds are men.
E. All Y is Z = All men are animals.
O. Some Z is not X = Some animals are not quadrupeds.

Converting the *major* premiss *simply*, and the *minor* premiss by *limitation*, as indicated by the *s* and *p*, we shall have

Ferio. Fig. I.

E. No X is Y = No men are quadrupeds.
I. Some Z is Y = Some animals are men.
O. Some Z is not X = Some animals are not quadrupeds.

It will be well for the student to reduce *every* imperfect mood, forming for himself particular examples under each.

Although we have made the subject of Reduction plain by the examples already given, we append a table of the manner of reducing each mood for reference, until the student is familiar with them. It is but a recapitulation in tabular form of what has been already explained.

Mood to be reduced.		Will reduce.	Process of Reduction.
Fig. II.	Cesare.	Celarent.	(s) Convert major premiss simply.
	Camestres.	Celarent.	(m) Transpose the premisses. (s & s) Convert the minor premiss and conclusion simply.
	Festino.	Ferio.	(s) Convert the major premiss simply.
	Fakoro.	Ferio.	(k) Convert the major premiss by negation.
Fig. III.	Darapti.	Darii.	(p) Convert the minor premiss by limitation.
	Disamis.	Darii.	(m) Transpose the premisses. (s & s) Convert the major premiss and conclusion simply.
	Datisi.	Darii.	(s) Convert the minor premiss simply.
	Felapton.	Ferio.	(p) Convert the minor premiss by limitation.
	Dokamo.	Darii.	(k) Convert the major premis by negation. (m) Transpose the premisses.
	Feriso.	Ferio.	(s) Convert the minor premiss simply.
Fig. IV.	Bramantip.	Barbara.	(m) Transpose the premisses. (p) Convert the conclusion by limitation.
	Camenes.	Celarent.	(m) Transpose the premisses. (s) Convert the conclusion simply.
	Dimaris.	Darii.	(m) Transpose the premisses. (s) Convert the conclusion simply.
	Fesapo.	Ferio.	(s) Convert the major premiss simply. (p) Convert the minor premiss by limitation.
	Fresison.	Ferio.	(s & s) Convert the major and minor premisses simply.

(41.) Indirect Reduction.

This process, called by the old logicians *Reductio ad impossible*, is analogous to the *reductio ad absurdum* of geometry. It consists in proving that the given conclusion *cannot be false* by proving, in *the first figure*, that its *contradictory* is false.

The symbols used to indicate the processes of *direct* reduction do not guide us in the *indirect* reduction, but we must deduce rules for this apart from the other.

To illustrate, let us take the mood

Fakoro. FIG. II.

A. All X is Y = All good men are virtuous.
O. Some Z is not Y = Some warriors are not virtuous.
O. Some Z is not X = Some warriors are not good.

If this conclusion *be not true*, its contradictory *All Z is X = All warriors are good, must be true*. Assuming this as true, and taking it in the place of the *minor* premiss in the syllogism, we shall have a new syllogism, as follows:

A. All X is Y = All good men are virtuous.
A. All Z is X = All warriors are good men.

from which premisses by our rules we draw the conclusion

A. All Z is Y = All warriors are virtuous.

But this conclusion must be *false*, because it is the *contradictory* of the original *minor* premiss, and the premisses were assumed to be true; hence one of these last premisses from which this conclusion is derived must be false; but it is not the *major*, for that was one of the *originally assumed* premisses; it must, therefore, be the *minor*, which we know to be the *contradictory* of our original conclusion; and the *original conclusion* must therefore be *true*: this, it will be observed, is proven in the first figure, in the mood *Barbara*. To take another example, let us reduce the mood

Darapti. Fig. III.

A. All Y is X = All gold is precious.
A. All Y is Z = All gold is a mineral.
I. Some Z is X = Some mineral is precious.

If this conclusion be not true, then must its contradictory,

No Z is X = No mineral is precious,

be so. Substituting this as the *major* premiss in the syllogism, we have

No Z is X = No mineral is precious.
All Y is Z = All gold is a mineral.

From which we draw the new conclusion

No Y is X = No gold is precious.

But this conclusion is false, because it is the contrary of the original *major* premiss, which we assume to be true; one of the premisses from which it was derived must be therefore false: it cannot be the *minor*, which was also assumed to be true; it must, therefore, be the *major*, which is the contradictory of the original conclusion; hence, the original conclusion must be *true*.

It will occur, in reducing many of the moods by this process, as in the last example, that we shall find the conclusion *false*, because it is the *contrary* and not the *contradictory* of one of the original premisses. By referring to the subject of *Opposition* (30), we see that if one *contrary* is *true* the other must be *false*.

Without presenting a greater number of examples of this kind of reduction, which the student may multiply for himself, we lay down the following rules for reducing the various imperfect moods.

Rules for Indirect Reduction.

1st. *In the second figure,* substitute the contradictory of the conclusion for the minor premiss, and proceed as above in the mood *Fakoro*.

2d. *In the third figure*, substitute the contradictory of the conclusion for the major premiss, and proceed as with the mood *Darapti*.

3d *In the fourth figure*, substitute the contradictory of the conclusion for the minor premiss, and proceed as before.*

As reference is always easier to a tabular form, we annex one showing in what perfect mood the indirect reduction of each imperfect mood will take place:

Fig. II.	*Fig.* III.	*Fig.* IV.
Cesare to Ferio.	Darapti to Celarent.	Bramantip to Celarent.
Camestres to Darii.	Disamis to Celarent.	Camenes to Darii.
Festino to Barbara.	Felapton to Barbara.	Dimaris to Celarent.
Fakoro to Barbara.	Datisi to Ferio.	Fesapo to Celarent.
	Dokamo to Barbara.	Fresison to Celarent.
	Feriso to Darii.	

Before proceeding to consider the irregular, informal and compound syllogisms, we pause to show the method of geometrical notation, already referred to, by which the pure syllogism may be expressed.

(42.) Notation of the Syllogism.

As there subsists in the mathematics such a relation of analysis to geometry, as that most analysis is capable of geometrical construction, and every form of geometry may be stated analytically in terms of its equation, so mathematical logicians have attempted to make for the analysis or symbolic form of the syllogism such a geometrical notation as shall at a glance represent to the eye, in areas of limited space, what the symbols do to the mind. Indeed, the idea is so simple that we have already illustrated the dictum of Aristotle through its agency. Many writers, however, have been inclined to go too far in its use.

* Except in cases of *Bramantip* and *Dimaris*, in which the contradictory is substituted for the major premiss, and the conclusion simply converted.

The schemes of notation best known are those of Euler, Ploucquet and Lambert, and the more complete one of Sir William Hamilton. This latter, however, passing beyond our needs, is suited to such changes as would result from the introduction of the *new analytic;* and, as we have advisedly declined to place that system in our text-book, it is sufficient to mention Sir W. Hamilton's scheme without explaining it. In a more extended historical treatise it would demand a special consideration. We can here only explain what we mean to use.

Euler's scheme of notation is altogether the one best suited to our purpose, and we shall limit ourselves to the explanation of that. It is essentially an arrangement of *three circles*, to represent the *three terms* of a syllogism, and, by their combination, the three propositions. Thus, if we have the judgment

All men are mortal,

we know that under this class—*all men*—are included many species and individuals; as, for example, *all Americans.* Representing, then, the sphere of the conception *mortal* by a circle, placing within this circle a smaller one, wholly contained in it, as the sphere of *all men*, and yet a smaller one, wholly contained in this latter, as the sphere of *all Americans*, we shall have—

which is the notation of a syllogism in BArbArA. By similarity of process, we shall represent the syllogism in CElArEnt:

INDIRECT REDUCTION. 115

No A is B,
All C is A,
No C is B.

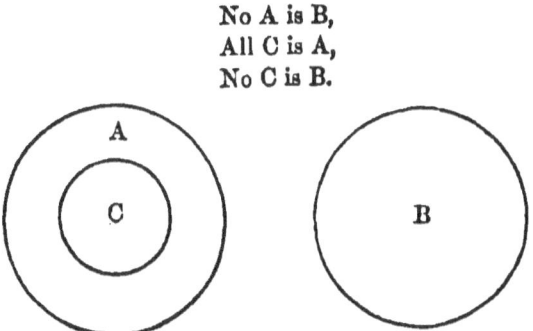

DArII will be thus expressed:

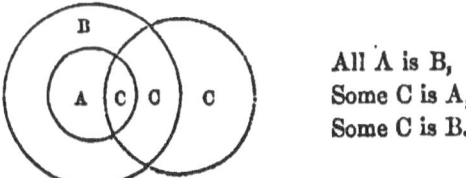

All A is B,
Some C is A,
Some C is B.

Here it is evident that it is only *that some C which is contained in A* that we have a right to assert is also contained in B, although other portions of C may by chance be also contained in B. FErIO:

No A is B,
Some C is A,
Some C is not B.

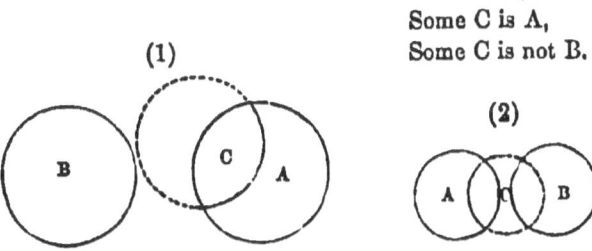

Here two cases are presented—where no C is B and where some C is B—neither of which affects the truth of the conclusion that *some C is not B*. We have only applied this scheme to the first figure, but by this simple notation of Euler every syllogism in the other figures may be represented to the eye, and made clear to those who are much quicker at geometry than at analytical work. Take for example *Darapti* of the third figure:

All A is B,
All A is C,
Some C is B.

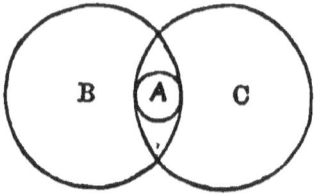

But besides this representation of valid syllogisms, this system exposes at once fallacious arguments and acts as a test upon a test of their unsoundness. Take for example the case of illicit process of the major term:

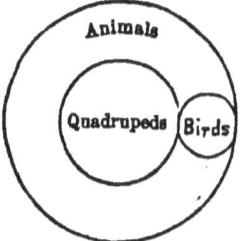

All quadrupeds are animals,
A bird is not a quadruped,
A bird is not an animal.

In which the figure denies the conclusion by allowing the premisses, and yet showing that *birds are* contained under the genus *animal.* Or if we take the case of the negative premisses:

No A is B,
No C is A,

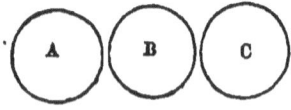

the figure shows us that there is no relation whatever established between or among the terms which would entitle us to a conclusion.

The student will find it easy and pleasant to write out all the moods and the logical fallacies by this circular method

INDIRECT REDUCTION. 117

of notation; and as two modes of coming at facts make the memory more tenacious of them, this practice will fix clearly in his mind the moods and figures of the syllogism.

The system also illustrates the categorical propositions as to the distribution of their terms, very satisfactorily:

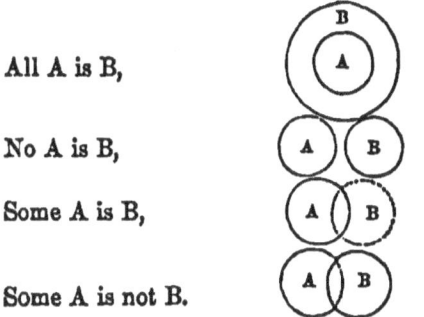

All A is B,

No A is B,

Some A is B,

Some A is not B.

It would be a good exercise for the student to be called upon to represent any given syllogisms by this notation.

CHAPTER IX.

OF IRREGULAR, INFORMAL AND COMPOUND ARGUMENTS.

(43.) **Of Abridged Syllogisms.**

WE have thus far considered only those arguments which appear directly and without analysis in the form of a simple syllogism, and have explained those processes which we perform upon known and acknowledged facts, stated as premisses and conclusion; but the mind of man sometimes passes intuitively over certain steps of these processes without stopping to express them, which gives rise to *abridged arguments;* or it halts in doubt and uncertainty, being not sure of its facts, but frequently balancing between two, one of which must be true, because of the truth or falsity of the other. This produces hypothetical syllogisms.

All these in the present chapter will be treated of as informal syllogisms, or arguments which are not syllogisms *in form*, but which, if they be valid, must be capable of being put into the syllogistic form.

The first of the abridged arguments to be considered, because the one in most common use, is

*The Enthymeme.**

The enthymeme is a syllogism with *one* premiss suppressed; it matters not which; thus, having the syllogism:

All men are mortal,
Cæsar is a man,
Cæsar is mortal,

* ενθυμεομαι, to conceive in the mind.

we may suppress the major premiss and write the enthymeme,

> Cæsar is a man.
> Therefore Cæsar is mortal.

Or, suppressing the minor premiss, we have,

> All men are mortal,
> Therefore Cæsar is mortal,

either of which is a satisfactory expression, *because all three terms of the syllogism are expressed* in either form of the enthymeme, and we can at once reconstruct the syllogism; thus, taking the *latter form*, with the *minor* premiss suppressed, we see by examining the conclusion, in which the major and minor terms are always contained, that *Cæsar* is the *minor*, being the subject of the conclusion, and *mortal* the *major*, being the predicate. *Men*, then, must be the middle term, and we at once compare it with the minor term to form the suppressed premiss; thus,

> Cæsar is a man.

By a similar process we may reconstruct the syllogism when the major premiss is suppressed.

It is worthy of observation that in ordinary discourse men suppress the *major* premiss habitually, as that to which the mind most readily yields assent, although, if the proof of its truth be required, the task would be more difficult than to establish the truth of the minor. Thus, in the example given above, we would take for granted as a fact that

> All men are mortal;

whereas, without the declarations of the Bible—and Logic, as a science, moves independently of any extraordinary or supernatural dicta—this proposition is incapable of proof; for, although all men have died thus far in the world's history, the process of induction cannot be finished until the end of man as a race.

But this seems like a cavil. The major premiss, although

thus incapable of mathematical proof, is the one which most surely demands belief; and so, when in the enthymeme we speak of the *suppressed* premiss, we mean the *major premiss*, unless it be otherwise explained.

As a simple rule for reconstructing the syllogism from the enthymeme, we observe that,

If the subject of the conclusion be found in the expressed premiss, that premiss is the *minor*. If the predicate of the conclusion be found in the expressed premiss, it is the *major*.

Sometimes it becomes necessary to put the enthymeme into logical form before proceeding to reconstruct it. Thus, the example given above might be, and most commonly is, thus spoken or written:

>Cæsar is mortal,
>*Because* Cæsar is a man;

which is evidently a transposed form of the enthymeme. Whenever the *causal* conjunction *because* unites the propositions of an enthymeme, we may invert the propositions and unite them with the *illative* conjunction *therefore*, and then proceed to reconstruct the syllogism; thus,

>Cæsar is a man,
>Therefore He is mortal.

Many abridged arguments which appear in a hypothetical form are in reality simple enthymemes; thus,

>If murder is a crime,
>The murderer should suffer.

In which there is really no hypothesis or condition in the premiss, because all allow that murder *is* a crime, and are consequently ready to declare that

>The murderer should suffer.

When the enthymeme has been reconstructed into a syllogism in any one of the figures, we shall be able to put it directly into the first figure, and can then apply to it the test of Aristotle's dictum.

THE SORITES OR CHAIN ARGUMENT.

(44.) The Sorites* or Chain Argument.†

The Sorites is an abridged argument consisting of a series of propositions in which the predicate of the first is the subject of the second, the predicate of the second the subject of the third, and so on until we combine the subject of the first and the predicate of the last to form a conclusion; thus,

A is B = The mind is a thinking substance.
B is C = A thinking substance is a spirit.
C is D = A spirit has no composition of parts.
D is E = (That which has) no composition of parts is indissoluble.
E is F = (That which is) indissoluble is immortal.

Concl. A is F = The mind is immortal.

This may be illustrated by a figure:

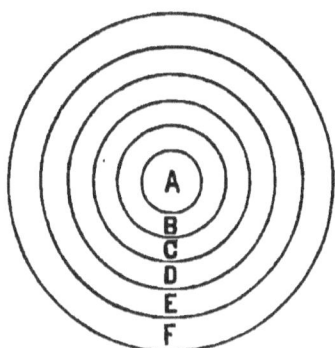

Now, if we try to put this collection of abridged arguments into the syllogistic form, in order to apply the dictum of Aristotle to them, we shall see that the Sorites is an abridgment of a series of syllogisms in the first figure; that the terms B, C, D and E, which are *used twice*, are middle terms, and that we may construct as many syllogisms as we have middle terms. Taking, then, the second proposition of the sorites, *B is C*, as the *major premiss* of the first syllogism,

---

* σωρείτης = a heap, or collection.
† Called by the Germans, more significantly, Kettenschluss, or chain argument.

and the first, *A is B*, as the minor, we shall have as a conclusion *A is C*, which we use as the *minor* premiss of a second syllogism, using the third proposition of the sorites as a major premiss; and so on, as long as the middle terms last; thus,

1st.	2d.	3d.	4th.
B is C,	C is D,	D is E,	E is F,
A is B,	A is C,	A is D,	A is E,
A is C.	A is D.	A is E.	A is F.

    A thinking substance is a spirit.
1st. The mind is a thinking substance.
    The mind is a spirit.

    A spirit has no composition of parts.
2d. The mind is a spirit.
    The mind has no composition of parts.

    That which has no composition of parts is indissoluble.
3d. The mind has no composition of parts.
    The mind is indissoluble.

    That which is indissoluble is immortal.
4th. The mind is indissoluble.
    The mind is immortal.

These are all in the first figure, and consequently are forms to which the dictum will directly apply.

It must be observed that in the sorites the first proposition, A is B, is the only one which may be *particular*, because it is the only *minor* premiss expressed, every other being used as a *major*, and we have already seen that in the first figure the *major* premiss must be *universal*.

So, again, the last proposition, *E is F*, is the only one that may be *negative*, for, if any other be negative, we should have in one of the syllogisms a negative *conclusion* which is to be in turn the *minor premiss* of the succeeding syllogism, and we have already shown that in the first figure the minor premiss must be *affirmative*. But the conclusion deduced from the last syllogism does not become a minor premiss, and so the last conclusion may be negative; it would then read thus:

No E is F.
All A is E.
No A is F.

Or the chain of the sorites would be broken in whatever place the negative proposition should occur.

The sorites is a very simple and conclusive abridged form of argument; for the mind, taking the only expressed minor term A, which is expressed in the chain, links it by jumping from middle term to middle term, B, C, D, E, to the final *major* term or F, as surely and more easily than in the syllogisms into which it is elaborated.

By its aid we easily establish the points in any great argument, either as recapitulating the process of the argument, or as stating them preparatory to a comprehensive discussion. Thus, to establish the effect of a republican government, we shall have—

The Americans make their own laws.
Those who make their own laws are free.
Those who are free are contented.
Those who are contented are happy.
Therefore The Americans are happy.

It is evident that the sorites may be properly stated in the inverse order, thus:

D is E, C is D, B is C, A is B,
Therefore A is E.

Here the sorites starts from its widest terms, D and E, to include the narrower and more limited terms, C, B, and finally A.

This form is called the *Goclenian Sorites*, from the name of its originator. It serves, perhaps, better to illustrate the fact stated that only the most *extensive* proposition, which in the ordinary form is the *last*, and in this the *first*, may be *negative;* which, as we have seen, will give us a negative conclusion, thus:

D is not E, C is D, B is C, A is B,
Therefore A is not E.

Hypothetical Sorites.

If we have a string of *conditional* propositions, such that the *consequent* of each becomes the *antecedent* of the *succeeding* one, the argument is called a *hypothetical sorites*, and the conclusion is obtained either by affirming the *first* antecedent with the *last* consequent, or by denying the *last* consequent with the *first* antecedent, thus:

1. If A is B, C is D; If C is D, E is F;
   But *A is B*,   Therefore *E is F*.

2. If A is B, C is D; If C is D, E is F;
   But *E is not F*,   Therefore *A is not B*.

Examples.

1.

If the Bible is from God, it should be taught;
If it should be taught, men should be set apart to teach;
If men should be set apart to teach, they should be supported;
But the *Bible is from God*, therefore its *teachers should be supported*.

2.

If the Bible is false, it deceives the world;
If it deceives the world, it should be destroyed;
But *it should not be destroyed*, therefore *it is not false*.

To the hypothetical sorites it is evident that the Goclenian form will also apply. Indeed, this is illustrated in the last case mentioned, where we reason back from the *denial* of the *last* consequent to the *denial* of the *first* antecedent.

(45.) Of the Epichirema.*

Most arguments employed in ordinary conversation and writing consist of simple syllogisms, abridged into enthymemes, linked together in a compound form; but in many cases the form of the syllogism is observed where the premisses are

---

* The Greeks seem to have considered this a great logical weapon, as the name they gave it signifies *a violent onset* or *laying of hands upon* ἐπι, and χείρ.

arguments in themselves. When the premisses are thus separately established, before the conclusion is deduced, the argument is called an Epichirema, thus:

> *The victors are injured by war, because it hardens their hearts;*
> *The French were victors at Marengo, for they retained the field;*
> *The French were injured by their victory.*

The *major* premiss is an enthymeme, which may be expanded into a syllogism; the same is true of the *minor;* hence we have two distinct arguments within the one which originally appeared. To apply the tests to their validity, they need only be written out in syllogistic form. In most apparently simple syllogisms there is in reality implied the epichirema. As for example, in the one given to illustrate the mood *Fakoro*, of the second figure,

> All true patriots are friends to religion,
> Some great statesmen are not friends to religion,
> Some great statesmen are not true patriots,

the major premiss demands in itself a reason, thus:

> *All true patriots are friends to religion,* because *religion is the basis of national prosperity and advancement.*

So also does the minor,

> *Some great statesmen are not friends to religion,* because *their own lives are not in accordance with its precepts.*

Each of the premisses given is an enthymeme; of which the clause *because, etc.* is the *premiss,* and the first statement, *all true patriots, etc.,* is the conclusion. Now, this *premiss* to the *premiss* is called the *prosyllogism.*

Sometimes the establishment of the final conclusion will warrant us in drawing other conclusions also, thus:

> A is B,
> C is A,
> Therefore C is B,
> Therefore X is Y, etc.

This *conclusion* from a *conclusion* (X is Y) is called the *epi-syllogism*.

In mathematics it is called a *corollary*, or something that flows from the demonstration without new proof.

To take the example before quoted, we shall have:

All true patriots are friends to religion.
Some great statesmen are not friends to religion.
Some great statesmen are not true patriots.
Therefore *They deceive their countrymen,*
and *Deserve no rewards from their country, etc.*

A number of syllogisms joined together in a connected argument constitutes a *Poly-syllogism*.

(46.) Of Hypothetical Syllogisms.

Corresponding to the various forms of hypothetical *propositions*—viz., conditional, causal, disjunctive, etc.—we have conditional, disjunctive and causal *syllogisms*. They are all of so simple a nature that the mind finds no difficulty in the ratiocination which they express; but as we have asserted that, if valid, they may be reduced to the form of a categorical syllogism in the first figure, we proceed to show how this may be done.

Conditional Syllogisms.

If we examine a conditional *proposition*, we shall see at once that the *affirmation* of the *consequent* will follow from the *affirmation* of the *antecedent;* thus:

If A is B, C is D = *If he has a fever, he is sick.*

But if we *deny* the *antecedent*, we may *not therefore deny* the *consequent*, since this consequent might spring from some other antecedent as well as from the one given, thus:

If A is not B = if he has not a fever,

we cannot say,

C is not D = he is not sick,

since

C might be D = he might be sick,

from some other cause than

A being B, or *his having a fever.*

For similar reasons we may pass from the *denial* of the *consequent* to the *denial* of the *antecedent,* but *not* from the *affirmation* of the *consequent* to the *affirmation* of the *antecedent.* When we pass from the *affirmation of the antecedent* to the *affirmation of the consequent,* the reasoning is called *constructive;* and when we pass from the *denial of the consequent* to the *denial of the antecedent,* it is called *destructive.*

We may have, then, *two, and only two,* forms of conditional syllogisms, *constructive* and *destructive.* To form the first, we take the *whole conditional proposition* as *the major premiss; the affirmation of the antecedent* for the *minor,* from which premisses we shall draw the *affirmation* of the *consequent* as the *conclusion,* thus :

Maj. prem. If A is B, C is D. = If he has a fever, he is sick.
Min. prem. A is B = He has a fever.
Conclusion. C is D = He is sick.

To frame the *destructive* conditional syllogism, we take the *whole proposition* as before for a *major* premiss, the *denial* of the *consequent* for a *minor,* and we deduce as a *conclusion* the *denial of the antecedent,* thus :

Maj. prem. If A is B, C is D = If he has a fever, he is sick.
Min. prem. C is not D = He is not sick.
Conclusion. A is not B = He has not a fever.

As these are the only possible forms of conditional syllogisms, and as we have shown that all other forms of hypothetical propositions—*disjunctive, causal,* etc.—may be easily reduced to conditional propositions, we have only to show how *these* conditional syllogisms may be reduced to the form of simple categorical syllogisms, and we shall, in effect, have shown it for all.

Considering first the constructive form, and remembering

that the *form* of condition may be removed by the phrases *"the case of"* and *"the present case,"* and that the proposition assumes the form of a categorical proposition, of which the *antecedent becomes the subject*, and the *consequent becomes a predicate*, we shall have for the *constructive form*,

	$\overbrace{\text{X}}$		$\overbrace{\text{Y}}$
Maj. prem.	The case of A being B	is	the case of C being D.
	$\overbrace{\text{Z}}$		$\overbrace{\text{X}}$
Min. prem.	The present case	is	the case of A being B.
	$\overbrace{\text{Z}}$		$\overbrace{\text{Y}}$
Concl.	The present case	is	the case of C being D.

Or, All X is Y. (A)
All Z is X. (A)
All Z is Y. (A)

which, X being the middle term, is evidently in the first figure, and the dictum may be at once applied. Using the same phraseology, and thus translating the *destructive form*, we have,

	$\overbrace{\text{X}}$		$\overbrace{\text{Y}}$
	The case of A being B	is	the case of C being D.
	$\overbrace{\text{Z}}$		$\overbrace{\text{Y}}$
	The present case	is not	the case of C being D.
	$\overbrace{\text{Z}}$		$\overbrace{\text{X}}$
	The present case	is not	the case of A being B.

Or, All X is Y. (A)
No Z is Y. (E)
No Z is X. (E)

which, *Y* being the middle term, is in the second figure, and in the mood *Camestres*, which must be reduced to the first figure or the form of the dictum.

If, now, we perform the operations indicated to reduce this mood (*m, s, s*), we simply convert the minor premiss, and then transpose the premisses, and simply convert the conclusion, we shall have,

$$\underbrace{\text{The case of C being D}}_{Y} \quad \text{is not} \quad \underbrace{\text{the present case.}}_{Z}$$

$$\underbrace{\text{The case of A being B}}_{X} \quad \text{is} \quad \underbrace{\text{the case of C being D.}}_{Y}$$

$$\underbrace{\text{The case of A being B}}_{X} \quad \text{is not} \quad \underbrace{\text{the present case.}}_{Z}$$

or simply converting the conclusion,

$$\underbrace{\text{The present case}}_{Z} \quad \text{is not} \quad \underbrace{\text{the case of A being B.}}_{X}$$

> No Y is Z. (E)
> All X is Y. (A)
> No X is Z. (E)
>
> or, No Z is X.

which is the form of *Celarent* in the first figure.

The logical form of the conditional does not depend upon the subject-matter of the propositions composing it. There may be, for example, two apparently independent propositions—that is, propositions in which the terms are entirely distinct—thus conjoined, or there may be a term the same in each, which will cause no difference in the *logical* form; thus we may have—

If A is B, C is D = If John remain, James will go; or,
If A is B, A is C = If the Bible is true, it (the Bible) deserves our attention.

To explain this apparent difference, it will be remembered that A, B, C, etc., although *terms* in the proposition, are not the terms of the syllogism when it is put in a categorical form, but that the *antecedent* and *consequent* become the true *terms*; and therefore it matters not whether there be *three* or *four* independent terms in the conditional proposition before its change of form.

A few examples of conditional syllogisms are given to accustom the student to the form, and to guard him against the improper use of it.

Examples.

1.

If the fourth commandment is obligatory upon us, we are bound to set apart one day in seven.
But the fourth commandment is obligatory upon us.
Therefore we are bound to set apart, etc.

2.

If any theory could be framed to explain the establishment of Christianity by human causes, such a theory would have been proposed before now.
But none has been proposed.
Therefore no such can be framed.

3.

If the eclipses of Jupiter's moons occur sixteen minutes later, when the earth is farthest from Jupiter, than when she is nearest to Jupiter, light must travel ninety-five millions of miles in eight minutes.
But these eclipses do occur so much later in the given position.
Therefore light travels at the rate stated, or, two hundred thousand miles in a second.

4.

If taste is uniform, all men will admire the same objects.
But all men do not admire the same objects (one sees beauty where another only finds deformity).
Therefore taste is not uniform.

Disjunctive Syllogisms.

A *disjunctive syllogism* is one the *major* premiss of which is a *disjunctive proposition* (26), and the *minor* a *categorical*.

Brutus was either a parricide or a patriot = Either A is B, or it is C.
He was not a parricide = A is not B.
He was a patriot = A is C.

Here, when the *major* premiss consists of two members only, the *minor asserts* the one and the *conclusion denies* the other; or, the *minor denies* the one and the *conclusion asserts* the other. Or we may have, instead of *two* alternatives, *three* or *more*, thus:

> The angle A must be equal to, or greater or less than, the angle B.
> But it is neither greater nor less than it.
> Therefore it is equal to it.

It is evident that the disjunctive syllogism may be at once stated in a categorical form by any simple phraseology which will rid us of the disjunctive form, thus:

> Brutus could not be at the same time a parricide and a patriot (but must be one of the two).
> He was a patriot,
> Therefore he was not a parricide.
> Or, He was not a parricide,
> Therefore he was a patriot.

Examples of Disjunctive Syllogisms.

1.
> It is either true that knowledge is useful, or that ignorance is so.
> But it is not true that ignorance is useful.
> Therefore knowledge is so.

2.
> Mohammed was either an enthusiast or an impostor.
> He was an enthusiast.
> Therefore he was not an impostor.

This is Gibbon's argument, but it is faulty in point of fact, for a man may be both enthusiast and impostor, and some men have a great enthusiasm for imposture.

3.
> A government either licenses a free press, or it is oppressive.
> The French government does not license a free press.
> Therefore it is oppressive.

4.
> A wise lawgiver must either recognize future rewards and punishments, or must appeal to an extraordinary Providence.
> Moses did not do the former.
> Therefore he must have done the latter.

Of the Dilemma, Trilemma, etc.*

A *dilemma* is a compound argument composed of *conditional* propositions upon which we reason *disjunctively*. When two conditional propositions are combined with a *disjunctive minor premiss*, the argument is called a *dilemma*. When three, four, etc. are so combined, they constitute a *trilemma, tessaralemma*, etc. The generic name *Dilemma*, however, is technically given to them all. Dilemmas are divided into four kinds, according to their being *simple* or *complex, constructive* or *destructive*.

A *simple* dilemma is one in which we have as a *major premiss* several antecedents with a single consequent, thus:

Maj. prem. { If A is B, If C is D, then X is Y, If E is F. }   *Min. prem.* { But either A is B or C is D or E is F }

Conclusion. Therefore X is Y.

A *complex* dilemma is one in which we have several antecedents, and each has its own consequent, thus:

Maj. prem. { If A is B, G is H. If C is D, I is K. If E is F, L is M. }   *Min. prem.* { Either A is B or C is D or E is F }

Conclusion. Therefore { Either G is H or I is K or L is M }

Now, if in the simple dilemma, instead of reasoning as we have done *constructively* from the *disjunctive affirmation* of the

* δις; τρεις, τεσσαρες, etc., and λημμα, from λαμβανω.

antecedents to the *disjunctive affirmation* of the *consequent*, we reason *destructively*—that is, *deny* the single *consequent*—then all the antecedents fall to the ground; there is no longer the condition of the dilemma; for we have a simple conditional syllogism. Or if we have *one antecedent* and *several consequents*, and reason *destructively*, it is as though we had but *one consequent*, since the *denial* of any *one* requires the *denial* of the *one antecedent*; thus, in the argument,

$$\text{If A is B,} \begin{cases} \text{C is D,} \\ \text{G is H,} \\ \text{L is M,} \end{cases}$$

it matters not whether we deny one or all the consequents, the denial of the antecedent follows. Hence, properly speaking, there is *no such thing as a simple destructive dilemma*. It differs in no wise from a simple destructive conditional syllogism.

The *destructive* dilemma proper, then, consists of *several antecedents, each* with its own *consequent*, in which we *disjunctively deny* the *consequents*—that is, deny any of them or all in turn—and we may *disjunctively deny* the *antecedents*.

Maj. prem. If A is B, C is D.   *Min. prem.* But either C is not D
         If G is H, L is M.   or   L is not M.
         etc.   etc.
         Conclusion. Therefore either A is not B,
                            or   G is not H.

To apply this abstract form to a particular example; let us take the argument of Antisthenes:

Maj. prem. If we conduct the affairs of state well, we offend men.
             If we conduct them ill, we offend the gods.

If now we reason constructively we shall add,

Min. prem. But, we must either conduct them well,
             or conduct them ill.
Conclusion. Therefore we must either offend men,
             or offend the gods.

If we reason destructively, we add, as a *minor premiss*,

But we must either *not* offend men, or *not* offend the gods,

and as a *conclusion*,

Therefore, we must either not conduct them well, or not conduct them ill.

To rid themselves of the perplexities of the dilemma, the old logicians always established from their premisses an undue, because not a logical, conclusion, but a moral and material one, a passage of the mind to a purpose which had been suggested by the matter of the argument; thus, the conclusion of Antisthenes from the perplexity of the dilemma was, that we had *better not meddle with the affairs of state at all.* Take another illustration:

> If a wife is beautiful, she excites jealousy;
> If she is ugly, she gives disgust;

and the illogical but common conclusion is

> It is best not to marry.

Most logicians have erred at the very outset by supposing that, because there is an alternative expressed in the dilemma, it is a *disjunctive* instead of a *conditional* syllogism, and thus have rendered it a vehicle of fallacy which it would be impossible for Logic to arrest; thus, they would read the last example,

> Either a wife excites jealousy by her beauty,
> Or disgust by her ugliness;
> Hence it is better not to marry.

In any such case, if we first put the dilemma in its true *conditional* form, and then (leaving the province of Logic, which presumes all given propositions to be true) examine the *subject-matter* of the propositions themselves, we shall find the falsity which causes perplexity; thus, it is not true *univer-*

sally, nor commonly, as is implied in the example, that *if a wife is beautiful she excites jealousy.* It is even *less true,* that is, in a fewer number of cases, that if *she is ugly she causes disgust;* hence the conclusion that it is best not to marry is *less true, i. e.,* applies to a fewer number of cases, than either of the foregoing assertions, *i. e.,* the falsehood is increased by the number of false statements preceding the conclusion.

It is evident that the dilemma may be resolved into as many conditional syllogisms as the greatest number of antecedents or consequents, and that these may be reduced according to the rules for the reduction of conditional syllogisms.

Any dilemma may also be stated in a categorical form. Thus,

> The case of A being B, is the case of G being H,
> The case of C being D, is the case of E being F;

and we may then proceed as in conditional syllogisms.

Examples of the Dilemma.

1.

If Eschines joined in the public rejoicings, he was inconsistent.
If he did not, he was unpatriotic.
But either he did join, or he did not.
Therefore, he was either inconsistent or unpatriotic.

The following dilemma was formed to confute the doctrine of Pyrrho, the skeptic, which was, that because everything has its contradictory, everything is *false;* or that no one could know anything certainly:

2.

If what you say is *true,* then there is something which is not false; *ergo,* your system is wrong.
If what you say is *false,* then it has no value as an argument; *i. e.,* your system is wrong.
But what you say must be either true or false.
Therefore, in either case your system is wrong.

3.

There are two kinds of things which we ought not to fret about—what we can help and what we cannot.

<small>(The student will put this in the form of a dilemma.)</small>

Having explained the various forms of argument, simple and compound, our next subject of investigation is of the erroneous use of these forms. To this has been given the generic title of *Fallacies.*

CHAPTER X.

FALLACIES.

(47.) The Meaning and Comprehension of a Fallacy.*

DIFFERENT terms are used to express the errors which are found in *terms, propositions or arguments* in Logic. Thus, we say of a *term*, when it is not *uni-vocal, i. e.,* when it has not *one* meaning and *only one*, that it is *equivocal* or *ambiguous, i. e.,* has more than one meaning; of a *proposition*, if it be not *true*, that it is *false*, which expresses in other words that the predicate and subject have no proper connection; of an *argument* we say, when it violates the dictum of Aristotle or any of the rules given, that it is *invalid*, and sometimes of an invalid argument we say that it is *fallacious*.

A *fallacy*, then, is *an invalid argument which appears at first sight to be valid*. If it be used *with the intention to deceive*, the fallacy is called a *sophism*.† An argument manifestly and foolishly invalid would then be neither a sophism nor a fallacy.

The subject of fallacies is one of the most important in the study of Logic, for not only is Logic designed to teach us to reason correctly, but also it should teach us to perceive and detect all errors in reasoning. Hence we find the earliest writers on Logic giving rules and cautions for avoiding and detecting fallacies.

The first division of fallacies which they have made is into

---

* *Fallo* = to deceive.

† Sophism comes through the word Σοφιστης, from σοφος, *wise*. Sophist was the name given in irony to those whose *wisdom* showed itself in an abuse of words and reasoning.

fallacies *in dictione* and *extra dictionem*. As *dictio* means the *form of words*, and not the *meaning of the words*, or what is expressed in our word *diction*, the class *in dictione*, or *fallacies in form*, will evidently come within the province of Logic, while those *extra dictionem*, not being in the form, but in the subject-matter, with which Logic is only indirectly concerned, will really not fall within the scope of our study.

But since the line between the two, although easy to be drawn, is continually mistaken in practical argument or controversy unless it be thus drawn, it becomes necessary to explain both classes with care, that we may always distinguish between the *truly Logical* and the *non-Logical* or *material* fallacies; and this is particularly important, because those who resort to fallacious reasoning use both these kinds of fallacy in combination with each other. One class of these material fallacies, which arises from the ambiguity in words, and is therefore called *verbal fallacies*, needs but a slight change, as we shall see, to become *formal* or *logical* fallacies.

(48.) Of Fallacies in Dictione, or Formal Fallacies.

These are the fallacies about which Logic is particularly concerned.

Under this class are included all violations of the dictum of Aristotle, and of the *axioms and rules* laid down for determining the validity of an argument. The fallacy in all cases under this head is apparent in the *form* of the expression; hence the name *formal* fallacies. Of this kind are—

1. Undistributed middle terms.
2. Illicit process of either term.
3. Negative premisses.
4. Affirmative conclusion from a negative premiss, and *vice versa*.
5. More than three terms in the argument.

Of these, repeated examples have been already given in

syllogistic form; it is only by putting them in this form that the fallacy is at once and easily detected.

But it should be borne in mind that in practice such fallacies are not stated in the *syllogistic* form, in which they are thus easily to be detected, but are stated in the form of an *enthymeme*, or other abridged argument, and so covered with words that the effect is produced without the mind being convinced—the conclusion allowed, because the mind cannot see the false steps which have been used, although it has not certified itself that the true have been taken. Let the student then take the trouble, in each such case, to write out the argument in syllogistic form, and, for greater clearness, *to use symbols*, and the invalidity will be apparent.

Thus, we are told that "a certain man was a good father, because he attended to the physical necessities of his children;" *food* and *clothing* and *shelter* being the criterion of a good father. Let us apply the test of Logic to such an argument:

Maj. prem. $\overbrace{\text{All good fathers}}^{X}$ $\overbrace{\text{provide for the physical wants of their children.}}^{Y}$

Min. prem. A B^{Z} $\overbrace{\text{did thus provide}}^{Y}$

Therefore A B^{Z} $\overbrace{\text{was a good father.}}^{X}$

Or, using symbols,
All X is Y,
Z is Y,
Z is X.

That is, Y, which is the *middle* term, is *undistributed*, being the predicate in two affirmative premises.

Again, it is asserted that "brutes are not accountable beings, because they are not responsible;" which involves a fallacy of *illicit process*. Thus,

		X		Y
Maj. prem.	All	responsible beings	are	accountable.
		Z		X
Min. prem.	Brutes	are not	responsible beings.
		Z		Y
Therefore	Brutes	are not	accountable.

All X is Y,
No Z is X,
No Z is Y.

In which *Y*, which is distributed in the *conclusion*—being the predicate of a negative proposition—is undistributed in the *major* premiss: *an illicit process of the major term.*

It will be observed, in this latter instance, that the *conclusion* is, we believe, a *true* one, but it is not reached by such premisses; and thus indeed it constantly happens, that men adopt a conclusion on internal grounds which they cannot explain, and then seek in every direction for premisses by which to substantiate it: and so, on the other hand, many a just statement loses credence, from the fact that weak and empirical men undertake to prove it by false premisses or fallacious reasoning.

It is further to be remarked that men who are guilty of fallacy in argument, either through design to deceive or weakness of reasoning power, are apt to combine many single arguments into a compound argument. If, then, one of these be faulty in its ratiocination, every ulterior conclusion is endangered, and the whole chain of argument is fallacious. To detect the error, therefore, requires that the whole chain be exposed link by link, and that the proper tests be applied to each argument. We have given examples of the fallacy of *undistributed middle* and *illicit process;* the student will not need illustrations of the other *formal* fallacies mentioned.

(49.) Material, or Informal Fallacies.

It will be allowed that in every fallacious argument the conclusion *does or does not follow from the premisses*. If it do not follow from the premisses, then when written out by symbols the fallacy is apparent, coming under one of the heads of formal fallacies which we have just enumerated. The fault here is evidently in the reasoning; but when the conclusion *does follow from the premisses*, when written out by symbols, the fallacy is not apparent, the fault will not lie *in the reasoning*, but *either* in the *premisses* or in the *conclusion, i. e.*, as to their *truth* or *falsity*, or as to the *ambiguous meaning* of words used in both. Such fallacies, with which Logic is not directly concerned, are called *Material* Fallacies.

It has been remarked before that Logic indeed takes for granted that the propositions composing its syllogisms are true, and that, when we write the general proposition A is B, no meanings shall be given to A and B which shall violate the truth of the proposition. If then we put for A, *Learning*, and for B, *useless*, and thus write,

Learning is useless,

or, by a change of words, the doctrine of the Stoics,

Pain is (a lesser sort of) pleasure,

we shall reason to *false* conclusions, *the matter of the propositions forming the syllogism being false*, while the *logic of the argument* may be *correct*. It must be allowed that *material* fallacies are more numerous and more fruitful causes of error than *the logical*, and as such deserve a special consideration, although indirectly allied to our subject.

We shall, therefore, endeavor briefly to give the principal forms or titles of material fallacies, and to illustrate them by examples, observing, at the outset, that they assume many and varied forms under these titles, all of which we cannot take the time to consider.

The simplest division of them is one which grows out of the consideration of—
1. *Errors in the premisses.*
2. *Errors in the conclusion.*

Of Errors in the Premisses.

Logicians have adopted technical names for the fallacies of this kind, viz.: the *petitio principii,* or *begging the question; Arguing in a circle; Non causa pro causa,* or *the assignment of a false or undue cause.* These branch out into various minor divisions.

As all these grow out of a *false* or *undue assumption of premisses,* they are akin to each other, and in many cases are not easily to be distinguished. Especially is this true of the first two.

I. *Petitio principii.* This consists in using as a premiss to support an adopted conclusion or assertion the same fact in other words. Thus we are told that "if the heart be touched death ensues, *because it is a vital part,*" or that "morphia produces sleep *because it is an anodyne.*"

Now what is it to say but that death ensues when the heart is touched, *because death doth ensue?* or that morphia produces sleep *because it produces sleep?*

Our language, which has so many synonyms from the Anglo-Saxon and the Latin, gives full play to this sort of fallacy, and many a wordy man is guilty of it without knowing his own error. And besides, this fallacy is the just recompense of those who endeavor to prove *axioms,* or who seek to penetrate into the ultimate facts for which God assigns no cause but the fiat of his own will.

II. *Arguing in a circle.* This fallacy depends upon finding a premiss to prove an asserted conclusion, and then, when asked for the proof of the truth of that premiss, endeavoring to make the conclusion prove the premiss; or, as this would be easy of detection, to make the circle still larger—*i. e.,* proving the truth of the premiss by a third proposition which depends upon the conclusion, and the playing upon these three, like the juggler's balls of which one is always in the air, but which, it is very difficult to tell. In case of the simplest form, writing out the syllogism will detect it; and in the latter and more complex case, the sorites, or its syllogisms written out, will find it out.

Thus, many men, not content with the everywhere shining proof within and without that there is a God, and mistaking the relations which the Holy Scriptures bear to him, would prove the *existence of a God from the truth of the Scriptures*, and then prove *the inspiration of the Scriptures* from the fact that *they came from God*.

> *As the Scriptures are the word of God, what they declare must be true.*
> *The Scriptures declare that God exists.*
> Therefore *That God exists is true.*

Or again:

> *The word of God must be true.*
> *The Scriptures are the word of God.*
> *The Scriptures are true.*

III. *Non causa pro causa.* This fallacy, which indeed may stand for the general title of unduly assumed premisses, consists technically in assigning as a reason or cause in the premisses one which has nothing to do with the conclusion, or one which is not itself proven, and is not therefore a sufficient cause. The first of these errors is called the fallacy of *a non tali causa pro tali*, or the assignment of a cause as though it were a cause, when it is not; and the second is the *a non vera pro vera*, in which the assumed premiss cannot be proven to be true as a cause, and may therefore be considered false. Under this head we have the fallacies technically called *post hoc ergo propter hoc*, or considering an event as a cause, because it precedes another event, and *cum hoc ergo propter hoc*, taking something for a cause when it occurs simultaneously with an event.

Of the latter of these divisions, the *a non vera*, we find a striking example, and an excellent logical retort, in the reported dialogue between Charles II. and Milton, after the poet had become blind. "Think you not," said the king, "that the crime which you committed against my father must have been very great, seeing that Heaven has seen fit to punish it by such a severe loss as that which you have sustained?" "Nay, sire," Milton replied, "if my crime *on that account* be adjudged great, how much greater must have been the criminality of your father, seeing that I have only lost my eyes, but he his head!" Another and common example of this is the following:

The natives of barbarous countries regard an eclipse as portentous of war and famine; and should they come together, they would assign it as the cause of their trouble. We know that it is not, but they only note the conjunction of the two as satisfactory proof that it is. Either of these may be easily written out in the syllogistic form, in which the propositions can be scrutinized as to their subject-matter and the falsity detected.

The fallacy of *a non tali* is chiefly used in analogous instances, where things which in one connection are useful or hurtful are assumed to be useful or hurtful in all; as because dry weather is good for the traveler it is also good for the farmer, or because the corn-laws were beneficial to England at one time they must always be so. Of the *a non tali*, the following example will serve as an illustration, viz.:

> All poisons should be avoided.
> Brandy and wine are poisons.
> Therefore They should be avoided.

That is, they are poisons only when taken in certain amounts and under certain circumstances. This is an invalid argument used by many good persons, the true reason for avoiding brandy and wine being *the danger of acquiring a habit of using them* to such an extent that they will be poisons.

Errors in the Conclusion.

We come now to the *second division* of material fallacies—those in which *the error lies in the conclusion;* they are all included under the general head of *Ignoratio elenchi, or irrelevant conclusion.*

The word *elenchus*, as used in the early writers, meant the contradictory of your opponent's assertion, and thus implies, what indeed was a feature in earlier Logic, the existence of an opponent. Dialectics were almost always in the form of dialogue, and the Socratic mode of questions and answers was adopted as the acutest method of argument.

The disputatious spirit of the Greeks was as much concerned about the victory in logomachy, or *word-war*, as about the discovery of truth, and hence arose many of their errors and paradoxes. This spirit of controversy and the constant keeping in sight of the *elenchus* has pervaded the methods of Logic to a very late period.

The *ignoratio elenchi* is the *ignorance of the contradictory of our opponent's assertion* which we display when, instead of establishing the *elenchus, i. e., proving the contradictory*, and thus proving *his conclusion* or assertion *false*, we attempt to establish something resembling the contradictory.

As it is not our purpose to reproduce the Grecian technicalities and method, let us get rid of this name and form, and call the fallacy, as it has been called by modern writers, the fallacy of *irrelevant conclusion.*

Those who employ it—and this, it may be remarked, is the most common and practical of all the material fallacies—generally state the conclusion as *a fact*, and when asked for the premisses or proof, are compelled to present such as display the irrelevancy of the conclusion. Thus, one asserts the fact that "Alfred the Great was a scholar," and when asked for proof says, "*Because he founded the University of Oxford.*" Now, there may be distinct proofs that he was a scholar, but this certainly is not conclusive. Let us state the syllogism:

> Those who found universities are patrons of learning;
> Alfred the Great founded the University of Oxford;
> Therefore *he was a scholar.*

The conclusion is irrelevant; the true conclusion being, from these premisses, that

> *He was a patron of learning.*

If polemical writings, and especially those which partake of the nature of popular and heated controversy, be analyzed, this will be found to be the standing fallacy, as often self-deceiving as deceiving others, and responsible for much of the widespread error in speculative science.

So varied is its nature that it has been from the early times known under various names and presents its insidious temptations to all kinds of persons.

Perhaps that form which is of most universal application is the *argumentum ad hominem*, the *unfair appeal to personal opinions, or to one's vanity or prejudice.* After exhausting all the arts to prove a thing wrong which is not so, the argument closes with "Well, *you* would not do so!" Even in matters of religion we are triumphed over by the adversary by a reference to ourselves and our own imperfect actions, when the question concerns the abstract truths of God's holy law. This form of the fallacy needs, then, a special watch as the most insidious.

Next in enumeration is the *argumentum ad populum,* which is the former fallacy extended from one individual to many, from personal opinion to popular prejudice.

Unprincipled demagogues use this fallacy continually; and where the sophistry would be apparent to any single mind gifted with common sense, the enthusiasm and thoughtless spirit of a mob, moved by a fiery harangue, is blind to its unreasonableness. This may be called the logic of revolutions.

A third kind of *irrelevant conclusion* is the *argumentum ad verecundiam*, or appeal to the *modesty* or sense of shame of our opponent, hoping that he will not presume *to attack respected authorities and time-honored customs*. It is based upon the general principle that natural prejudice is in favor of the existing and the old. Although healthful progress may have demonstrated their errors and provided us with better methods, the cry is of recreancy to our fathers' memories, to old associations, to History; and thus the world has been trammeled and clogged by what professes to be the genius of conservatism, but what is in reality the genius of obstinate error.

The *argumentum ad superstitionem* is an appeal to one's superstition, from which, in some form or other, few men are free; *ad odium* is to one's hatred; *ad invidentiam*, to envy; *ad amicitiam*, to friendship. Many others might be formed following this analogy. Those mentioned are sufficient to illustrate the principle.

Besides these forms of *irrelevant* conclusion, there are many which have been proposed in pleasantry, such as the *argumentum ad baculinum*, and others which Sterne humorously refers to in "Tristram Shandy."

There are, however, it must be particularly observed, many cases in which many of these arguments are not fallacies—in which, indeed, they may with great propriety be used, clothed with all the graces of rhetoric and imbued with all the spirit of enthusiasm.

The *argumentum ad hominem* is not a fallacy when the design is to teach pure truth, and when no *unholy passion* or emotion of man is appealed to. In this application it was used by our Saviour himself to the Jews on many occasions with great force and beauty. His touching and yet searching appeal to them for the woman taken in adultery sent them out *one by one* before its power. Each one felt the argument and admitted the conclusion.

His arguments in favor of *healing on the Sabbath*, and *searching the Scriptures*, that they might find every page luminous with Him whom they denied, were examples of the unfallacious and powerful use of this form of reasoning.

So, too, an appeal (*ad populum*), not to the prejudices, but to the conscientious scruples and feelings, of a multitude, is without fallacy, and is productive of the best results.

Many customs, long honored and dear to every heart— customs national, civic, professional, domestic—unmingled with error, unopposed to progress, make the *argumentum ad verecundiam* a most proper and effective appeal.

But such is the waywardness of man that the temptation to fallacy in their use is exceedingly strong, and must be carefully guarded.

Argumentum Ad Rem and Ad Judicium.

Opposed to all these, when used as fallacies, are two forms of valid argument: the first expresses a concentration solely upon the reason of the *thing itself*, and is therefore called the *argumentum ad rem;* the second is when the appeal is made to the unbiased exercise of the individual *judgment;* this argument is called *argumentum ad judicium*. Many writers have increased the number of these fallacious *argumenta* to a much greater extent; but those given are the principal ones, and will sufficiently indicate the process by which they are coined when needed.

Changing the point in dispute.

Another form of the "irrelevant conclusion" is the fallacy of *changing the point in dispute*, in which one of the parties in a long and difficult controversy, after having tried in vain to establish his irrelevant conclusion, dextrously shifts his ground from the point in dispute to some other, and pertinaciously claims that to be true which *has not been disputed*, while the true matter of contention is left without an honest confession of his inability to prove his assertion. For example, a person undertakes to prove that the people in general are not educated: *i. e., he first denies that they are;* but failing of this, he really proves, what no one denies, viz.: that *all the people should be educated.*

Fallacy of Objections.

It has been remarked that Ignorance may state in a few words objections against Science which wise men could not refute in whole volumes. The truth of this is manifest. The error of reasoning from the statement or existence of these objections to the falsity of the science is one of the forms of irrelevant conclusion which has been called *the Fallacy of Objections.* It consists in asserting that, *since there are objections against a Science, that Science is false;* whereas the judgment demands that the claims of the Science as well as the objections be duly stated, and that the turning of the scale decide whether truth or error predominate. If it be a complicated system, it will be found to contain portions of both; if an abstract theory, it will stand or fall by such a test. This fallacy has been industriously aimed by skeptics against the mysteries of the Christian faith, but it soon loses its point in such an encounter.

From the consideration of the various species of the fallacy of *irrelevant conclusion* which have been mentioned, and the examples given, it will be seen that it is in all its forms the standing sophism in houses of legislative convocation—that it is the demon of debate. Few subjects of debate are so abstract and unit-like but that dull minds will find room to wander about, one losing the very point in question, another concerned about a crowd of details which have little or no bearing upon it, a third mistaking the fine and delicate points of the logical argument; some, becoming heated in the controversy, will lose their temper and reasoning powers together, and, overpowered by the truth and Logic of their opponents, will have recourse to appeals to the prejudices and interests of their audience; and others, more shrewd than just, will seek to bring by similar means the cause and persons of their adversaries into disrepute by the light arrows of ridicule or the more ponderous weapons of insult. It is amidst such scenes, and under such circumstances, that the master mind shows itself as it rises over the storm of the debate, and brings them back first to the consideration of the subject in dispute in its true and abstract form. Perhaps the most striking illustration of this is found in our own Congressional history. After Mr. Webster's first speech on "Foote's resolution," many senators had delivered their views, and much sectional excitement was aroused. Mr. Webster began his famous second speech, with just such a master-effort to come back to the true merits of the controversy:

"Mr. President, when the mariner has been tossed for many days in thick weather and on an unknown sea, he naturally avails himself of the first pause in the storm, the earliest glance of the sun, to take his latitude, and ascertain how far the elements have driven him from his true course. Let us imitate this prudence, and before we float farther on the waves of this debate, refer to the point from which we departed, that we may at least be able to conjecture where we now are. I ask for the reading of the resolution before the Senate."

The resolution was read; the Senate found their true position, and Mr. Webster's speech is as masterly for its logic as for its oratory.

(50.) Verbal Fallacies.

There is still a most important class of invalid arguments to be considered; it is that growing out of the *ambiguous* or *equivocal* meanings of words, many words being identically the same, and yet bearing widely different meanings. Thus, the simple word *line*, when used in different connections, means many distinct things: for example, *a cord used in fishing; a few words in a letter; an arrangement of troops or ships in battle array;* and when we see the word *porter*, we are in

VERBAL FALLACIES. 151

doubt which of three meanings is intended—a *gate or door-keeper*, a man who *bears burdens* or a *kind of malt drink*.

In most such cases, however, there is a single root to which we may trace all these secondary meanings; thus all the meanings of a *line* refer to the mathematical definition that it is *length, without breadth or thickness*, and all the uses of *porter* refer to the Latin word which signifies *to bear*.

It is true that there are examples of words spelt alike which have different etymologies, but these are few: *host* from *hostis*, and *host* from *hostia* in the sacrifice of the mass, are examples of this; so also *league* from *ligare*, to bind, and *league* from the Latin *locus* or distance between *places*, contracted in French to *lieue*, as the word *focus* is into *feu*, are examples of such words. With these few illustrations of ambiguous terms, let us see how they are used in argument.

The ambiguous word is *sometimes the middle term* and *sometimes it is the major or minor;* in most cases, however, it assumes the former place, so that the general name given to this form of verbal fallacy is "the Ambiguous middle."

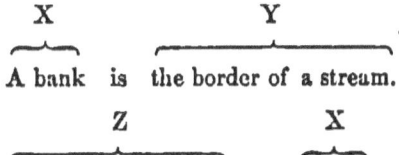

               X   Y
     A bank is the border of a stream.
           Z   X
     This stone building is a bank.
Therefore This stone building is the border of a stream, etc.

Now, if this glaring and absurd fallacy be stated by symbols, we shall have—

          X is Y,
          Z is X,
          Z is Y,

which is the form of a valid argument in the first figure; so that the fault lies in the matter of the propositions which compose the argument, and not in the form, which is correct; the fallacy then must be classed, with such an investigation, among the *material* and not among the *formal* fallacies. But let us go a step farther; since "a bank" in the major premiss means something entirely different from "a bank" in the minor, they are in reality different terms; let us symbolize them by different letters, and calling the first X, let us call the second P; we shall have, writing by symbols, as before,

$$X \text{ is } Y,$$
$$Z \text{ is } P,$$
$$Z \text{ is } Y,$$

a *formal* fallacy, in which there are, contrary to the rules laid down, *four* terms instead of *three;* and this comes within the province of Logic. The fallacy of *Ambiguous middle* has very justly, then, been called by logicians a *semi-logical fallacy; before we discern the ambiguity* it is a *material* fallacy, with which Logic is not concerned; but *as soon as we discover the ambiguity*, it discloses *four* terms which make it a *formal or logical fallacy.* It is because of this peculiarity, and because it is so very much used in common life, that we treat of it under the distinct head of *verbal fallacies.* But we have said that it is not only in the *middle* term that this ambiguity occurs; it also happens in the *major* and *minor* terms, and is quite as sophistic when it lurks there as in the middle term. We have therefore discarded the title "Ambiguous middle," as applied to the general class, preferring "Verbal fallacies," as more truly illustrative of the error in any of the terms.

There are many ways in which words come to be used ambiguously, and we shall give a few of them, with illustrations; and first we place the influence of *Etymology.*

I. *Etymology.*

A word which originally meant one thing now means quite another, and the fallacy consists in using it in the *two senses*, in two propositions of the syllogism. Thus, taking the first meaning of *pagan* to be a villager (paganus*), and its present

* From *pagus*, a village.

meaning to be a believer in some other religion than that of Christ, we have—

> A *pagan* is a disbeliever in Christ;
> Every villager is a *pagan;*
> Every villager is a disbeliever in Christ.

Akin to this, and indeed ranging under the general subject of *etymology*, is the use of *paronyms*, or *paronymous words*.

Paronymous words are the noun substantive, adjective, verb, etc., belonging to each other and springing from the same root. *To project, próject, projection, projector*, etc. are paronyms, springing from the Latin compound of *pro* and *jaceo*. So *presume* (in its two senses), *presumption, presumptive, presumptuous*, etc. are paronyms growing from the root *presumo*.

Take the following example, in which the ambiguity will lie in the middle term:

> *Presumption* is impertinence;
> That the sun shines, *I presume* (or, is *my presumption*);
> Therefore I am impertinent (*in asserting that the sun shines*).

It will be remembered that the true logical form of the minor premiss, which is usually written, "I presume that the sun shines," is—

subj.   pred.
That the sun shines   is presumed by me.

Again:

> To propose a railroad is a *project* (or a *projector's* work).
> This man proposed a railroad.
> Therefore He is a *projector* (or visionary man).

In which the ambiguity lies in the major term. Now, no one can work advisedly without making *projects*, whereas one of the meanings of *projector* is a *scheming* and *visionary* man who ought not to be relied upon.

II. *Fallacy of Interrogations.*

This is a use of two or more terms in a question, making thus in reality two questions, requiring two distinct answers, and the ambiguity lies in the *single* answer given to both. It is common for those who use this fallacy to express but one question, while the other is implied. Thus, if a man who has always been temperate is asked, "*When he gave up drinking?*" the implied question is, "*Did he ever drink?*" and then, *if so, when did he cease?* or, in the celebrated inquiry of King Charles II., "*Why a live fish does not add to the weight of a vessel of water?*" the implied question being "*Does a live fish add?*" etc., and if so, "why?" etc., or a witness may be asked, Where were you when the prisoner murdered the deceased? which would imply what remains to be proved, viz., that he did murder him. This fallacy, which is called by the writers *Fallacia plurium interrogationum*, is made more subtle by the number and closeness of resemblance of the points included in the questions.

III. *Amphibolous Sentences.*

Sometimes the ambiguity, instead of residing in the words which compose the argument, lies in the construction, and thus, by different punctuations, we have double and opposite meanings. This passes from the *ambiguous* words to *amphibolous* sentences. Among the most celebrated of these is the response of the Delphic oracle to Pyrrhus when he went to encounter the Romans:

> Aio te Æacida Romanos vincere posse,
> Ibis redibis nunquam in bello peribis.

In the first line, either accusative may be taken with the infinitive, thus making either "Pyrrhus" or "the Romans" *able to conquer;* and in the second, *nunquam* may qualify either *redibis* or *peribis.*

So also in the Nicene Creed, we have, in reference to our Saviour, the words, "being of one substance with the Father, by whom all things were made."

The latter clause, so manifestly introduced by the Council

to declare the creative power and Godhead of *Christ*, in reality by strict rhetoric applies to "*the Father.*"

The name given to this fallacy is the fallacy of *amphibolous*[*] sentences, *i. e.*, tossed from one to another with a doubtful meaning.

Causes of Ambiguity.

Having mentioned the various kinds of ambiguity in words, we come to consider *why words have two or more meanings*.

We have already seen that many words expressing simple primitive ideas grow by usage to have other meanings, in which, however, the primitive idea is to some extent retained; thus, *line*, in all its meanings, adheres to the mathematical notion of *extension in length*.

Now, without being able to trace the exact process in all cases by which a word is thus gradually changed, we find that it ranges itself under one of these heads : 1. *Resemblance;* 2. *Analogy;* 3. *Association;* 4. *Ellipsis;* 5. *Accident.*

1. *Resemblance.* Many things bear the same name from their actual similarity in appearance. Thus, in carpentry, a *dove-tailed joint* is so called from its similarity to a *dove's tail*, or a *spear of grass* from its resemblance to the military weapon, a *spear*. So in the military art a "priest-cap" or "swallow-tail" is a redoubt so named from its actual resemblance to one of these two things, and a "crow's foot" takes its name from the form of a bird's talons.

2. *Analogy.* Our ordinary speech is full of the use of this figure of speech, and this fact has contributed to the ambiguity in many words. As resemblance is a similarity in appearance, *analogy* is a *similarity in use, purpose or relation*. Thus, we speak of the arm of a chair, because it holds the relation to the chair which the arm does to the human body; and thus an arm-chair is a chair which has arms.

We speak equally of a *sweet* food, or a *sweet* sound, because

[*] αμφι and βαλλω.

there is a similarity between the relations of the food to the palate and the sound to the ear. So a sour lemon and a sour individual create relatively similar effects upon the taste and upon the mind.

Ambiguity of resemblance and of analogy are both produced and perpetuated by the use of metaphor and comparison, in our ordinary discourse, and a wayward fancy, expressing itself in the social exaggerations of the day, is robbing some of our best words of their true shades of meaning; for example, *sweet, lovely, horrid, agony, wretch,* are deflected from their original meanings entirely.

An argument from analogy may lead to probability, but is fallacious when it claims a *certain* condition, but it may well be used to corroborate and strengthen other arguments as an additional likelihood.

3. *Association.* By this we mean the connection of parts in the same structure or institution, or to produce a single result. Thus, a *door* is the opening in the wall or the swinging shutter that closes it. *Faith* is belief, and *"the Faith"* is the system of Christianity. *Shot* is the leaden pellet: *a good shot* is either the person who shoots or the effect of the shot.

It is by the association of ideas, which, unlike our examples, are subtle and difficult to fix and determine, that fallacies have grown out of this ambiguity; and such is the want of correctness in the language of the great number of people that the tendency to this fallacy of words, expressing associated ideas, is particularly strong and dangerous.

4. *Ellipsis.* Another habit into which men naturally fall, in trying to avoid the use of many words, and words conveying thoughts which the mind will readily supply without their being expressed, is the use of *elliptical language.* While in most cases this is harmless and even profitable, in some it leads to error. Thus, we speak constantly of Scott, Byron, etc., when we mean their *works* or their *persons.* We use the form "to my father's," "at Mrs. Smith's," when we mean the

houses or "parties" of these persons, and such ellipsis is always understood; but many persons are deceived in their business relations by such ellipsis as the statement of another's wealth at so many thousands of dollars, when in reality, although it may produce the interest on such a sum, it cannot be made available for anything like the amount of the principal sum mentioned.

5. *Accident.* It seems in certain cases as though a word had assumed two meanings in a manner inexplicable and accidental. Such, for example, is the word *light*, which is equally opposed to *heavy* and *dark*, and which in conduct means the opposite of *serious* or *dignified*. But even in such a case we shall find one idea, however subtle, pervading them all, and that is the removal of a covering of some sort; thus, light removes the pall or covering of darkness; the incumbent weight of something heavy; the just restraints of dignity and sobriety. In strict truth, then, there is no *accidental* ambiguity, for, although there may be words in the double meanings of which we can discover no relation to a single idea, that relation undoubtedly exists, and by a profound research the number of such words would be very much diminished.

Many words are forced into a double meaning by a popular or political use, which may be called *accidental*, but which in reality is designed by one party as an *equivoque*, or stratagem, in the way of retort upon the other. It was thus with the use made of the word *Pretender* by the English Jacobites. When it became treasonable in any way to maintain the claims of James Stuart, the son of James II., who was called "the Pretender," they toasted him in the well-known verses:

> God bless the King; God bless the Faith's Defender;
> God bless—no harm in blessing—the Pretender.
> But which is the Pretender? which the king?
> God bless us all—that's quite a different thing.

It is evident that such a use of the word would deceive no one; nor was it indeed so designed, but rather to violate the spirit and yet adhere to the letter of the law. The true argument used by the adherents of the new dynasty was—

Those who aid a pretender to the English throne deserve punishment.
James Stuart is a pretender.
Those who aid James Stuart deserve punishment.

It must be understood that pretender in both premisses has the same meaning—*i. e., false claimant.*

But there is still another form of ambiguity which leads to fallacious arguments; it is where the ambiguity lies not in words, but in the *context;* or where our assertion means one thing when taken in a general sense, and quite another if considered in a special sense. Of these fallacies, arising from ambiguity in the context, there are two kinds:

1. *The fallacy of accidents.*
2. *The fallacy of division and composition.*

Under the first head are included the *Fallacia accidentis,* and the *Fallacia a dicto secundum quid ad dictum simpliciter.* These are the converse of each other.

Fallacia accidentis.

This is where, in one premiss, we assert something of a subject in a general sense, and, in the other, place upon that subject some *accidental* peculiarity which will lead us to error in the conclusion, thus:

Things bought in market we eat.
Raw meat is a *thing bought in market.*
Therefore Raw meat is what we eat.

Here the middle term is *things bought in market,* and it is considered in the major premiss as to its *essence,* viz.: that these things are in market for *general* use as food; in the minor we lose sight of its *essence,* and only regard some *accident* of it, viz.: that the meat bought in market is *raw.* Thus, in reality, the error is thrown upon the middle term, which is shown to *be not one,* but *two distinct terms,* and the fallacy is thus exposed.

The other form of this, which for shortness is called the Fallacy of *Quid*, may be translated *reasoning from the restricted or limited sense of a term* (secundum quid—*i. e., ali quid* in the monkish Latin), *to its broad or unrestricted use* (ad dictum simpliciter). Thus:

> This man is *innocent* (of a certain crime);
> But the *innocent* (entirely) are sure of Heaven;
> Therefore This man is sure of Heaven.

Fallacy of Division and Composition.

In this fallacy the middle term is used in its *collective* or *additive* sense in one premiss, and in its *distributive* sense in the other. When the middle term is used *collectively* in the *major* premiss, and *distributively* in the *minor*, the fallacy is of "Division;" when the *reverse takes place*, it is a fallacy of "Composition." The following are examples:

> *Fallacy of Division.*
> The Christians (as a sect) were persecuted at Rome.
> Constantine was a Christian (individually).
> Therefore He was persecuted at Rome.

> *Fallacy of Composition.*
> Three and two are two numbers (distributively).
> Five is three and two (additively).
> Five is two numbers.

Positive and Negative Intention.

Akin to these fallacies are those absurd conclusions reached by a play upon certain negative words, such as *nothing* and *no*, when used as an adjective; thus,

> Nothing is better than Heaven.
> A shilling is better than nothing.
> Therefore A shilling is better than Heaven.

> No cat has two tails.
> Every cat has one tail more than no cat.
> Every cat has three tails.

In these examples the middle terms *nothing* and *no cat* are taken in a *positive* sense in the *major* premiss, as though they expressed *living* or *existing* things, while in reality they mean *non-existence*. In the minor premiss they are taken in their true *negative* sense.

The best method of refuting them is to deny the major premiss, or to demand that it be put in other words, thus:

> It is not true of *anything* that it is better than Heaven;

which will foil the one who wishes to draw the absurd conclusion. It should be observed that such arguments are really used only in sport, but it is well to detect and understand the error which they contain.

(51.) The Manner of Removing Ambiguity in Terms.

The true method of ridding ourselves of this ambiguity of terms in argument is to demand a *definition* in each case, and to keep our *terms distinct* when thus defined. It will not, in most cases, be necessary to give a *real* definition, as a *nominal* one will answer every purpose. The ambiguity is usually such that by giving the true, limited and exact *name* (which is the province of a nominal definition) we shall detect and remove it.

In many cases where the fallacies consist of a number of arguments and many ambiguous terms, the first thing to be done is to disentangle the web of sophistry by writing them out in full and in due order, and then, after detecting the terms in which the ambiguity lies, to demand a definition in a few but plain and conclusive words in every case.

The equivocal nature of the word becomes apparent if we change the language, as in the translation of the familiar example into Latin—

> Light is contrary to darkness,
> Feathers are light,
> Therefore Feathers are contrary to darkness,

we shall have—

> *Lux* est contraria tenebris.
> Plumæ sunt *leves*.
> Plumæ sunt contrariæ tenebris.*

---

* Latham's Logic, p. 221.

This change of language, it will be seen, is of the nature of a definition.

(52.) The Fallacy of Probabilities, or the Calculation of Chances.

This consists in stating two *probable* premisses, and then drawing a *certain* or more probable conclusion, as though the number of probabilities combined amount to certainty, whereas, in most cases, the conclusion will be less probable than either, thus:

> Those who have the plague *probably* die;
> This man *probably* has the plague;
> Therefore He will (*certainly*) die.

Whereas, suppose *ten* out of *twelve* of those who have the plague die, then if we express *certainty* by the number 1, that probability is expressed by the fraction $\frac{10}{12}$ or $\frac{5}{6}$; and if it is an even chance whether or not he has the plague, that probability will be expressed by $\frac{1}{2}$. The probability of the conclusion, therefore, will be $\frac{5}{6} \times \frac{1}{2} = \frac{5}{12}$, or as $\frac{1}{2}$ is the expression for perfect *doubt*, *i. e.*, an even chance of his living or dying, he is less likely to die than to live, his chances of dying being 5 out of 12, and of living, 7 out of 12.

This fallacy is practically used in times of sickness and mortality, when fears of evil, excited by nervousness, affection, etc., place an anticipated conclusion for the true one.

When, instead of one syllogism or enthymeme, many are combined to make a compound argument, and the errors of probability are thus multiplied, the result will be at once farther from the truth and more difficult to detect.

Let us deduce then a simple rule for the calculation of probabilities. The subject has been called "the doctrine of chances."

When we speak of *chance*, we really mean *probable results of God's laws*, and in the use of either word we express our ignorance of the connection between natural causes and effects

Now, as that ignorance may be partial or entire, we are thrown upon a calculation of the chances, and we shall find that the probability ranges between the two extremes, *certainty* and *impossibility*. We do not pretend to assert by this that man may divine the results of God's doings in the future; but that, according to the action of natural laws and the sequence of an established order, we may approximate to the truth without assuring ourselves of it.

Thus, in throwing dice, we cannot be sure that any single face or combination of faces will appear; but if, in very many throws, some particular face has not appeared, the chances of its coming up are stronger and stronger, until they approach very near to certainty. It must come; and as each throw is made and it fails to appear, the certainty of its coming draws nearer and nearer.

The probability of a single event depends upon the number of chances of which it is one. Thus, if A is in a single action where 10 men are killed, his company numbering 50, the chance which each man stands of being killed, and consequently that of A, is $\frac{10}{50}$ or $\frac{1}{5}$. If we subtract $\frac{1}{5}$ from 1, or *certainty*, we shall have $\frac{4}{5}$ for his chance of being saved. The calculation of probabilities becomes more complicated where the events are combined. Thus, if in a second action 10 men more are killed, his chance of being killed *in this last action* is as 10 to 40, or $\frac{1}{4}$, and that of his being saved $\frac{3}{4}$. If now we would determine his chance of being saved, after both actions, we must multiply the two chances together: $\frac{4}{5} \times \frac{3}{4} = \frac{12}{20} = \frac{3}{5}$, which is as it should be, since 20 men are lost of the original 50 and 30 remain; his chance of being among the latter should be as 30 to 50, or $\frac{3}{5}$.

It is upon this principle of calculating chances that insurance companies are founded, and it finds a benevolent issue and scope particularly in those life-assurance companies which, demanding but a small percentage, making a large aggregate, are thus enabled to pay to widows and orphans an honorable support, snatching out of the jaws of death the means of life and social comfort.

It is, however, upon a false study, or rather in an ignorant and fatal reliance upon this principle, that those who frequent gaming-houses throw away their means, reputation and life; for the true gainers are not the frequenters of the gaming-table, but the keepers, who are acting upon this very doctrine of chances. By a calculation of chances it is found that, *in the long run*, the keeper of a gaming-house must win in almost every kind of game played, while only an occasional player, with what is called a *marvelous run of luck*, chances to win largely.

The subject of probabilities, which in its right use is not fallacious, but is reduced to arithmetical accuracy, has been placed under the general head of *Fallacies*, because of its being so liable to fallacious use, and so much employed thus. Mingling as it does with the superstition in our nature, we deem those things more probable than they are which we *desire* or *fear*.

The wish is father to the thought for *pleasant* hopes, and presentiments of evil are taken for its probable coming in our *gloomy* periods. We give a rule by the use of which all this may be avoided.

Rule.—The probability of any event is expressed by a fraction of which the numerator is the number of chances in its favor, and the denominator is the sum of all the chances; and the probability of any two or more events jointly occurring will be obtained by multiplying together the fractions expressing the probability of each.

(53.) Popular Fallacies.

It will be well, before closing the chapter on Fallacies, to show their practical use, especially in a popular illustration. A community, a state, a nation, will unite upon a fallacy from which it will be a sort of social treason to dissent; an age will be tinctured by error, pervading all classes, which only the innovation of a succeeding age can remove; a false principle will cling to human nature, in the mass, during many centuries, which the philosophic mind can only deplore in secret.

It will be our purpose, then, to put forth some of the simplest forms of popular fallacy, beginning with the most general. Some of these have been already mentioned in their logical places, as the different forms of *irrelevant conclusion*, etc.

I. The fallacy which is expressed by the adage, *Nil de mortuis nisi bonum.* There is a just meaning to this indeed; it is that the tongue of private enmity should be silenced—that we should consider Death as having adjusted all difficulties as between man and man, and awed our mortal infirmities into a silence and forgetfulness of the evil which existed in him who is now dead. So far the adage is good; but when it becomes a principle in public morals, when it tinctures the historian and the historical biographer, who should deal with the dead as with living defendants, arraigned for trial, its evil nature is apparent. When it eulogizes the dead at the expense of the living, and runs riot in obsequious praises and flattering epitaphs, it assumes its most sophistic form. "The same man," says Jeremy Bentham, "who bepraises you when dead would have plagued you without mercy when living." The reason of this is apparent. A dead man cannot be a rival; he incurs nobody's envy, and is removed from all the results of malice.

II. Not unlike the preceding is the fallacy conveyed in the trite saying, *De gustibus non est disputandum.* This is used fallaciously to put a stop to controversy; the assertion implying that as God gave man each his own taste, *one taste* is as good as another. But all our systems of education teach us that this is not true—that there is, on every subject which comes under the dictum of taste, a true standard which can and ought to be used. It certainly is better to put an end to controversy by saying that it is better to differ than to become excited and quarrel, than *falsely* to state that there can be *no dispute about tastes.*

III. There is a fallacy which particularly assails patriotism: it is the fallacy of asserting *that any one form or system of government is abstractly the best.* The Russian deems that men cannot be controlled in masses without single autocratic power; the Englishman defies the world to pick a flaw in his limited monarchy and superb aristocracy; while the American boldly declares that the best government is the democratic representative form. Where such men as Milton and Locke have "astonished the world by signal absurdities" in their models of government, we might be sure that its theory must be difficult; but the truth is, there is no abstract theory of human government.

Asiatic barbarians, when they leave their patriarchal wandering life, as in Russia, and come into the first corruptions of a half-civilized life, *must be governed by despotic power:* they cannot be republican; while on the other hand, it is only where education is general among the people—that they may know their wants, and how to supply them, and where individual honesty and virtue are everywhere felt, that no undue means may be taken to bring about such an end—that a democratic government is the right one. Then, in this freest form there is a reciprocal influence between the government and that upon which it is founded. A free government *enlightens* and *purifies* the people, while the enlightenment and purity of the people strengthen and ensure the government under which they live.

IV. There is a popular fallacy which may be called *Sweeping classifications.* It consists in ascribing to an individual something really belonging to another individual, only because the two happen to be of the same class; thus, during the French Revolution, when the fate of Louis XVI. seemed to hang upon a thread, one pamphlet was issued with the title

'The Crimes of Kings." Now, as there had been many bad kings in Europe and not a few in France, Louis XVI., the best of them, was put into the category of condemnation simply because he was a king.

Thus misusing the adage "*ab uno disce omnes*," governments and institutions, both secular and religious, are blamed because some of their members indulge in crimes entirely their own. The entire body is made to share in the condemnation because the few are guilty.

V. Space would fail in which to enumerate the current and manifest popular fallacies, most of which are used in legislatures and councils, and are considered in the light of shrewd and dextrous diplomacy. There is the "*no precedent* argument." It is stated thus: "The plan proposed is entirely new. This is certainly the first time such an idea has been broached in this honorable house; and *therefore* the secret hope is that this house will not now entertain it."

Next, we have personalities introduced, laudatory or abusive, by which to turn the current of the argument.

Another form is the assertion with regard to any measure that as "no complaint has ever been brought against it before, it must be a good one."

But perhaps the most insinuating form of popular fallacy is that of *authority* by which a man is required to join one or the other party in every question, thus causing the young ignorantly and prematurely to commit themselves to views and measures which later experience teaches them to be wrong; if *then* they change they are *traitors* or *turncoats*, if it be a national or political question, and *fickle* and *unreliable*, if it be of a less general nature. It is lamentable to see party guides bringing those under their control forward to swell the ranks of their party, and those thus introduced glorying in their new distinction, when self-interest and not truth has been the motive on both sides.

CHAPTER XI.

THE FUNDAMENTAL LAWS OF THOUGHT, OR FIRST PRINCIPLES OF REASON.

HAVING thus explained the various logical processes by which the mind seeks to establish truth and detect error, and having explained the subject of *fallacies* in form and in matter, the existence and prevalence of which show the necessity of an exact system of logic, it will now be proper to lay down for students the fundamental laws of thought, or what may be called the *first principles of reason.*

A *primary* principle is one which has no cause or reason behind it upon which it depends. It is recognized as true without proof, for it cannot be proved; it is necessary, universal and underivable—that is, it belongs to mind as a necessary part of its existence, it belongs to all minds, it depends on nothing antecedent of itself.

The number of these first principles has been more or less extended by different schools of philosophy, but there are *four* upon which most philosophers are agreed, viz.: IDENTITY, CONTRADICTION, EXCLUDED MIDDLE and the law of REASON AND CONSEQUENT. Upon these as a basis the system of Logic is reared as a superstructure.

I. IDENTITY. With the belief or cognition of our own existence comes the belief that *whatever* is, *is*, or, in the language of the older schoolmen, *Omne ens est ens.* In regard to any object the mind at once affirms it of itself, and cannot think of it but as existing. The law of identity, it will be readily observed, is the principle upon which logical affirmative propositions and definitions are formed. Thus, in the proposition All A is B, the identity of the whole of A with a part of B is set forth.

II. CONTRADICTION. Simultaneously with this intuitive belief in identity appears the second principle, *Contradiction*, or, in the words of Sir William Hamilton, more properly *non-contradiction*, which has been called the *highest of all logical laws*, which gives sole value to identity. The law of contradiction declares that we cannot conceive of a thing as *being* and *not being* at the same time. If identity declares that *A is A*, the mind refuses its assent to the contradictory, *A is not A*. Upon the law of contradiction is based all negative judgments and logical distinctions.

III. EXCLUDED MIDDLE. This law asserts that there can be no medium between the dictum of identity and that of contradiction, or it excludes such a medium. The two propositions, *A is A* and *A is not A*, being of such contradictory nature that if one is true the other must be false and *vice versa*, no medium between them can be conceived. We must think of either the one or the other as existing, and they cannot co-exist. The law of excluded middle, it will have been seen, has been set forth in a disjunctive proposition, *Either A is A or A is not A*. By identity and contradiction we conclude that if one contradictory proposition is *true* the other is *false*. By the action of excluded middle we reason from the falsehood of one to the truth of the other.

IV. REASON AND CONSEQUENT. The principle here set forth has been called also that of SUFFICIENT REASON. This implies that wherever a reason exists there must exist a consequent, and conversely, wherever we have a consequent, there must exist a sufficient reason for it.

Logic applies this principle directly in the reasoning process, and forms in close and necessary connection the series of notions which thought has produced. The axiom of Reason and Consequent must be kept quite distinct from that of Causality, as will be seen.

THE FUNDAMENTAL LAWS OF THOUGHT. 169

It is a significant fact that these laws were not developed by philosophers in the order stated. The principle of *contradiction* was enounced by Plato and emphatically stated by Aristotle, while the law of identity was not enounced as a coordinate principle until long after. Hence there has been a controversy among philosophers which of these is the first or highest principle; some assert that our own existence even is not a primary datum of intelligence, but is an inference from the existence of thoughts and feelings of which we are immediately conscious. Some would claim Identity to be first in order, while others regard Contradiction as the principle by which Identity is established, and without which it cannot be. So too there have been those who doubted whether contradiction was really a primary principle; *an a priori datum* of intelligence, or whether it was not a generalization from our earliest experience. With most, the essential *fact* is identity, the essential *law*, contradiction. Leaving such matters to the metaphysician, we may not only agree to consider contradiction a primary principle, but go farther, and assert that it lies, as it were, at the foundation of the others, and is implied in them. It is clear; it is universal; it is necessary.

It is *clear*, as is shown by the fact that it depends on the same evidence as the simple notion of existence, of which it is an affirmation, in that whatever *is* cannot *not be*. Thus it establishes *identity*.

It is *universal*, because, as the idea of being is implied in every apprehension and in every principle, so is this distinct affirmation of it applicable to all.

It is *necessary*, because by it reason must be guided in all its judgments, since through the *excluded medium* it establishes the absolute truth that *being* and *not being* cannot subsist together.

Let it be observed that we do not say that the other principles may be *demonstrated* by the principle of contradiction, but it holds place as the highest principle only as the others may be resolved into it. Demonstration supposes the thing to be demonstrated less evident than *the medium quo*, that by which it is demonstrated. Now, each of these first principles has its own intrinsic and immediate value and truth, and cannot be demonstrated.

It further appears that the law of contradiction governs all the principles of reason as a motive and guiding power, influencing the intellect to give its assent to that which without it would be incomplete and inert.

In *necessary* truth, the intellect affirms the truth of the principles which it perceives, because it sees the necessary connection between the two ideas compared, and at once explains or rather satisfies itself of the necessity by the principle of contradiction; or the truth of this principle, as an intuitive, undemonstrable truth, is sanctioned by the truth that its contradictory cannot be.

And what is seen in necessary truth is equally manifest in *contingent* truth. Truth is *contingent* when it depends for its existence upon some *hypothesis* or *condition* or cause or fact. Here the mind discerns that the truth exists because the condition exists and not otherwise, and hence by the law of contradiction that its contradictory must, on the same condition, be false.

Without entering into the speculations of philosophers in all ages of history, it seems to us that the principle of contradiction is the foundation and ultimate reason of all proof and of all assent of the intellect; that, so to speak, it gives vitality to the law of *Identity*, and suggests the necessity of *excluded medium*, establishing itself as a *sufficient reason* for both.

These principles are intuitive cognitions or *à priori* conviction, perceptions from which we reason; concerning which we cannot. To this extent they are incomprehensible; we know them to be, but not how and why they are. They are called à priori principles because they are *before* all our experience and before all possibility of proof.

Upon them are based, with greater or less claim to intuitive judgments, numerous axioms, such as *the whole is greater than a part, and a part less than the whole. Two things which are equal to the same thing are equal* to each other. By extension of these axioms in Logic, we have also *two terms which agree with one and the same third, agree with each other; and of two terms, if one agree and the other disagree with the same third, they will disagree with each other.*

It will not be without interest to say a few words in this connection concerning the question of *causality*, or, whence do we derive our notion of cause and effect? Various solutions have been proposed, so diverse and conflicting as to be in themselves properly named *series implexa causarum.* It seems to lie so near the first efforts of the mind that many philosophers have supposed the judgment of causality to be an à priori knowledge referable to a special principle of intelligence designed for it, and it alone. Others have variously considered it to result from experience, induction, generalization and custom.

In common language, the phenomenon may be thus stated: we cannot think of anything beginning to be without thinking of its having already existed in another form—that is, the necessity of our intelligence makes us believe of anything that it has a cause; and as the cause by the same process is believed to have a cause, the mind of necessity goes back in the chain of causes until it reaches the idea of a first cause. What is the limit of the mind in this search? This limit has been called *The Conditioned*, and the law of the conditioned is, that all that is conceivable—as, for example, in time and space—is bounded or limited by extremes which are inconceivable and contradictory, one of which must therefore be true. Thus we have, as one set of inconceivable extremes, absolute commencement and infinite non-commencement, both of which are inconceivable, and yet one of which is true. *The conditioned* is based upon the principle of contradiction, and it explains the true theory of causality. Thus the judgment of causality is a derived judgment, not from the power of the mind, but from its impotence to attain to the extreme. When an object appears to us as commencing to be, we cannot but suppose that what it now contains has existed before in some form—that every thing we see is an effect which must have had a cause—but why, or, of what, the cause is, we may be, and in some cases must be, ignorant. This inability of the mind to reach final causes, and thus to complete the explication of the principle, is expressed in negative adjectives, *infinite, unending, illimitable.*

CHAPTER XII.

(54.) Of Certain Modes in which Logic is Applied.

It is not within the scope of this work to enter upon the subject of *applied Logic:* this would require an investigation of all the sciences, or at least of a very numerous classification; but it is designed to explain the meanings of certain phrases which refer to the general applications of Logic.

We have the phrase *moral reasoning,* and it is often used as if conveying an opposite or contrary meaning to *demonstrative reasoning.*

This has reference, not, as we have clearly shown, to the kind of reasoning, as there is but one, but to the nature of the evidence employed, the meaning of *evidence* being *that testimony which sets forth the truth of a proposition.* Then, *moral reasoning* is the use of evidence in *moral* subjects, and *demonstrative* reasoning its use in mathematical subjects.

Now, *evidence* may be of *three* kinds—that is, as to the manner in which we obtain it; it may be *intuitive, inductive* or *deductive.*

Of Intuition, Induction and Deduction.

We come now to consider the means of discovering truth which are most useful, but which have been strangely confounded with Logic. They are processes as much bound by logical laws as all other movements of the reason are.

It is evident that, in order to the logical process, we must have *premisses;* now, these premisses are obtained evidently by the three methods just mentioned, intuition, deduction and induction or experiment.

By *intuition* we mean the immediate and absolute knowledge which, without any apparent effort, we find implanted in us. Such, for example, is the aspiration of man's soul after a Deity, as exemplified in the religious systems of all people, even the most barbarous, and such as the existence of certain affections and notions of moral conduct. In brief, consciousness in most of its forms and *the testimony of our external senses* are said to be sources of *intuition*. The truth of axioms is dependent upon the laws of identity and contradiction.

But most of our knowledge is derived from what we possess already in another form, as where we deduce certain inferences from acknowledged premisses or from observation and experiment, and generally many observations or experiments are necessary before we can determine a general law; thus, it required centuries of observation to determine the Copernican theory of our solar system; and almost all the developments in natural science are the fruit of many observations and experiments aggregated in each case to form one general law. It is an effort of man by a close study of the *phenomena* (φαινομενα) or appearances of nature, to arrive at some degree of acquaintance with the *noumena* (νουυμενα) or essences of its objects.

To unite these was the aim even of the heathen philosophers, and with their obscure lights they worked ardently in the labor; it remained for a doubter (Sextus Empiricus), two centuries after the coming of Christianity, to connect them for another purpose, and that was to arrive at a suspension of all judgment on objects whose nature is obscure, and thus to acquire a certain repose of mind (αταραξια) and perfect equanimity of disposition (μετριοπαθεια). But the inductions of Sextus were never really performed; he theorized to his skepticism, and his theories will not bear the rude hand of physical practice.

In order to illustrate the difference between *induction* and *deduction,* let us suppose a law already determined, which we state in the proposition *A is B.* Let any number of particular examples, as x, y, z, range under this law, thus, x is A, y is A, z is A, and we can manifestly reach the conclusion that x, y and z are all and severally B.

But suppose the general law unknown, and that it be approximated to in proportion to the number of particular examples; we shall thus have x is B, y is B, z is B, etc.; but x, y, z, etc., as we increase the number of the examples, represent the class A; hence we may state the law A is B, the truth of which will depend upon the number and extent of the experiments performed and particular instances observed. Or, to recapitulate in syllogistic form:

Deduction.	*Induction.*
(*Law*) A is B.	(*Part. examples*) x, y, z, etc., are B.
(*Part. examples*) x, y, z, etc., are A.	A is the class to which x, y, z, etc., belong.
(*Conclusion*) x, y, z, etc., are B.	(*Law*) A is (likely to be) B.

Now, there are certain sciences in which, from the nature of things, we can never state more certain results from induction than this *likelihood;* but this likelihood, it must be observed, becomes greater and greater, and at length touches absolute certainty, when we examine many particular instances and find none of them failing to range itself under the law which we call *likely,* so that at the last we write it to all intents and purposes as a categorical proposition, *A is E.* In some sciences we may exhaust all the particular examples and finish our induction by a certain law; or if by induction we find any quality or property to belong to the essence of the object undergoing the experiment, induction in both cases has led, as the other could not, to certainty.

There are two kinds of induction, *material* and *formal;* and it is by a want of proper distinction between them that the error has arisen of comparing induction improperly with the syllogism, and asserting that while induction is one kind of reasoning the syllogism is another—*i. e.,* deduction.

Hence, Lord Bacon and his followers, finding that *deduction* generally moved from what was contained in known premisses to lower classes or individuals contained in them, threw aside the syllogism as useless, and inaugurated *induction* as the new Logic of experimental philosophy. A simple examination of material and formal induction will set us right. Material induction is the process of experiment and observation—the laborious investigation of facts as to their discovery and their combination—but formal induction is obtained by the use of the syllogism itself, not confined, as some writers have attempted to show, to the third figure, but in most examples capable of being at once written out in the first figure, the form in which they may be immediately tested by the dictum of Aristotle, as in the example:

Maj. prem. Whatever is true of the cow, goat, deer, etc., is likely to be true of all horned animals;
Min. prem. Rumination is true of the cow, the deer, etc.;
Concl. (Law). Rumination is *likely to be* true of all horned animals.

The naturalist receives this as the only just conclusion from the formal induction to which the syllogism has helped him; but having as yet found no exception to the rule, he writes it out boldly and without fear of contradiction,

All horned animals are ruminant.

Of certain modes of using Syllogisms.

Argument à priori.—This is the mode of passing from known antecedents to necessary consequents, or, in the sciences, from *cause* to *effect*. Thus, if we consider the being of a God and of his attributes to be independently known, as by intuition, then we reason *à priori* to the existence of his works, the universality of his providence and the gracious designs of his redemption; this reasoning is most plainly stated in the form of the constructive conditional syllogism, the affirmation of the *antecedent*, or cause, helping us to the affirmation of the *consequent*, or effect.

Argument à posteriori.—This is reasoning from *effect to*

cause. If, by an inverse process, we first study natural religion, and experiment upon the wonders of the human mind and then pass back from these works around us to the establishment of the existence of a first great cause who *must have made them* all, we are said to reason à *posteriori*, or from results to their causes.

Of the two modes of reasoning, both are useful and effective, but the reasoning à *priori* is the most explicit, stating at once the cause and reason of the effect and conclusion, whilst that à *posteriori*, though equally conclusive, is not so explicit, because it simply proves that the conclusion must be true, although not stating its intrinsic cause. Thus, we prove the existence of a first great cause from his works, or à *posteriori*, since he is self-existent and therefore has no cause, and consequently his existence cannot be proven à *priori*.

History uses both forms, and combines them with great success. Taking, for example, on the one hand, the early elements of a nation's life—its people, its geography, its tendencies of government—history seeks to trace these to their legitimate *results* among the changing scenes of national existence; while on the other, looking around at the present condition and conduct of a nation, she takes these results, and tracing them back, in careful combination, with each step removed from the present, she seeks for their early and prime *causes* in the classic times of the country's origin.

Argument à fortiori.—This is a method by which we establish a *stronger* conclusion even than ordinary premisses need to warrant us. Thus:

> A is greater than B.
> B is greater than C.
> A is greater than C.

That this conclusion is just there can be no doubt, and that the form of it is not exactly that of the regular syllogism is equally apparent.

To apply the doctrine, let us present the argument by geometrical notation, and we shall have—

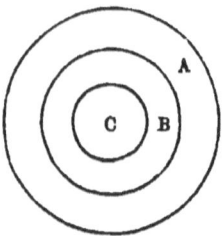

in which we have the relative greatness of A, B and C.

But we are entitled, it is evident, to put this in the syllogistic form:

B is A,
C is B,
Therefore, à fortiori, C is A,

which is *Barbara;* or ordinarily Barbara is itself the argument *à fortiori*, and is only otherwise when A, B and C, instead of being unequal, are exactly coincident.

In this latter case, we have the old case of *convertible* terms in *each* proposition, which is not set forth in separate form by the Aristotelian Logic.

This reasoning *à fortiori* is very effective and proper, and was used by our Saviour in his invectives upon Chorazin, Bethsaida and Capernaum with thrilling effect. So also is it forcibly used by the apostle to the Hebrews (x. 28) in the words: "He who despised Moses' law, died without mercy under two or three witnesses: of how much sorer punishment shall he be thought worthy, who hath trodden under foot the Son of God," etc.

The Investigation and Discovery of Truth.

We shall now briefly notice the forms of method used in investigating and discovering truth, to which at every step the canons of Logic may be applied.

Man has an inherent desire to find truth, and the universe around and within, the realms of nature and the domain of mind call that desire into constant activity. This curiosity, which "grows by what it feeds on," leads at once to the discovery of truth, and through the process to the education and development of his faculties.

The methods and order of investigation are: 1. Observation; 2. Supposition or hypothesis; 3. Induction; 4. Theory; 5. Fixed law or fact.

I. *Observation* is applied in general to whatever is presented by the senses; by it man discerns at once objects and facts. It includes a thoughtful, attentive outlook upon creation and a determination by the senses of the marked distinctions between existing things, and leads to the next step in the order of inquiry.

II. *Hypothesis* or *supposition*. Because certain things or conditions exist, we suppose the existence of causes which produced them, or of certain determinate effects which spring from them. The words hypothesis (Greek, $\dot{\upsilon}\pi o \tau \iota \theta \eta \nu a \iota$) and supposition (Latin, *sub* and *pono*) have the same meaning—an *underlying* basis upon which to build. A hypothesis assigns a probable cause or a reasonable connection. It is indeed a gratuitous assumption, but it has been so subjected to metaphysical conditions that it may be correctly and profitably used in our process of investigation. A just hypothesis is one which explains many phenomena and contradicts none, and it is a necessary condition that it should do so when no other hypothesis can. Thus it establishes a presumption in favor of our law or conclusion which must stand or fall by the next step in the process, Induction. Hypothesis is often incorrectly confounded with theory.

III. *Induction* is systematic experiment, based upon hypothesis. Having made a *nidus* for our observations and experiments, we test it by phenomena; if they agree with or range themselves under the hypothesis, we approach a general law; or if not, we see that the hypothesis is wrong, and assume a new one.

IV. *Theory* is the probable establishment of our hypothesis through the medium of Induction. In proportion as our experiments conform to the hypothesis it becomes probably true; as the experiments increase in number and still conform the probability approaches certainty, until at length we either reach certainty or, satisfied by sufficient induction, assume it, and arrive at a fixed law or fact.

It will be obvious that these forms of method, although distinct, really run into each other more or less as we proceed; that in preliminary observation we may use the simpler modes of induction; that in hypothesis we are anticipating theory, and hoping that we have probably assumed a law which shall be arrived at. But in systematic investigation they are mainly used in this order. Thus, Franklin *observed* the similarity between the spark from an electrical machine and a flash of lightning; he *supposed* or assumed as a hypothesis that they were the same; he *experimented* by flying his kite and leading the lightning along its string, and he then stated his *theory* of electricity, which, covering all phenomena and contradicting none, has assumed the character of an established law.

Of the Nature and Kinds of Evidence.

As the investigation of truth, according to the methods just stated, depends on evidence, we shall merely state the nature and kinds of evidence by which truth is established. Evidence (*e* and *video*) is that which makes a fact or proposition clear and obvious to the mental vision.

Consciousness, the *knowledge of the existence* of the thinking subject, comprehends all its phenomena; *intuition* is the act of the mind by which it looks at and into itself; through it we have a belief in our own existence, faith in the testimony of our senses, a reliance upon the uniformity of Nature's laws.

Sensation is the effect produced upon our senses by contact with the world around us. Sensation does not separate the object producing it from ourselves. Perception is the impression made by an object upon the mind through sensation. Through perception we gain the idea of *outness* or externality, and thus detach the object from ourselves. These are the conditions necessary to evidence, and to these must be added *memory* in its most extensive meaning, as the conservative, the reproductive and the representative faculty of the mind through which these conditions are made available.

Analogy is that resemblance between circumstances, relations or effects of two objects by which the mind is led to accept what is true of one as true of the other; as evidence it is by no means sure, but often corroborative, where other evidence is produced. *Induction*, or systematic experiment, is valuable as evidence; in the words of Bacon, "Prudent questioning is half the science."

And last we have the *Testimony* of mankind, which is based upon our natural inclination to believe in the experience and truth of others. It is evident that Testimony will depend for its value upon the capacity, the character, the prejudices, the means of knowing and the number of the witnesses.

An individual of average mind has the capacity to understand the moral and physical circumstances by which he is surrounded. The natural desire of man is to speak the truth, and all men unite in despising a liar; and while on a given subject one man may have only partial knowledge, many who are cognizant of it, by bringing each his own partial knowledge, will present fuller and more trustworthy testimony than any one by himself. Where *fact* is in question, the truth may thus be readily obtained; for the chief requisite is honesty: where *opinion* is desired, we must add superior knowledge and aptitude which will give authority.

CHAPTER XIII.

A HISTORICAL SKETCH OF LOGIC.

(55.) Division of the Subject.

HAVING completed, in general outline, the study of the formal Logic, in its present condition of exactness and practical use, we are ready to go back to its feeble beginnings, and trace it in its slow and trammeled movements from the days of the early Greek Philosophy, through the applications of Roman Science, the enlightening process of Christianity, the era of the scholastic subtleties, the dawn and advance of Experimental philosophy and the metaphysics of the eighteenth century, down to the controversies of our own day.

Nor are we yet to regard the science of Logic as established beyond dispute, and fairly stationed among its sister sciences; it is yet an arena of dispute, and the most distinguished philosophers disagree, as has been seen, even as to what it is and as to what is its scope.

It would be of great interest and profit to take such a historical view in detail, but the limits of this work will not permit it, and, besides, for all practical purposes, the periods of the history naturally divide themselves into four. These so much transcend all others in interest and value, and so absorb the events which just precede or immediately follow them respectively, that they form the plainest and most convenient method in which to present the history of Logic. They may be marked by the titles—

1. Aristotle.
2. Christianity and Logic.
3. Bacon, and the rise of Inductive Science.
4. The present system.

1. Under the first may be classed all the efforts of the human mind in the arrangement of a canon of reasoning, in that early time when knowledge, preceding method, was only seeking in darkness and obscurity that system of laws and principles by which alone knowledge may be made available. Around Aristotle, too, cluster the great expansions of science which were due to the conquests of Alexander and the great kingdoms of his successors.

2. In the coming of Christianity, Logic found, not a rival, but a guide, and in the early Church it was the weapon of their spiritual warfare. To the Church, as the representative of Christianity, is due much of the good of scholasticism.

3. Logic was the servant, the ill-used servant, of Inductive philosophy, and owes much of its long bondage and oppression to the illustrious founder of the system of Experimental philosophy.

From these considerations it has been assumed that we are better able to look into this history now that we are acquainted with the scope of the science; otherwise, we might fall into the same error, by reason of the honorable company in which we should find ourselves.

4. Since the time of Lord Bacon, and perhaps by reason of his example in condemning the syllogism, Logic has been degraded from its position as the controller of the reason on all subjects, and has been so intermixed with Mental philosophy as quite to lose its identity and be miscalled by its own name. This was its condition during the eighteenth century. In the nineteenth there have sprung up many champions of Aristotle and the syllogism, among whom first in distinction is Archbishop Whately. The universal principle of reasoning has been rescued by him from oblivion and degradation, and Logical science, although still maligned and fiercely attacked, seems ready to take its permanent place among the great elementary sciences of human investigation and instruction.

(56.) Aristotle.

It must be considered that the progress of such a science as Logic was necessarily gradual and slow; that from the beginning men had been contemplating the operations of the reason, or were making vain but progressive efforts to distinguish the exact functions of the reason, among the lazy elements of the human intellect. Many men had collected much material which lay floating in a chaotic state upon the great deep of the human mind.

The logical doctrines of *conception* as expressed in *terms*, of *judgments* as formed in *propositions*, were known to Socrates and Plato. Indeed, Zeno the Eleatic, who is mentioned as the inventor of Dialectic, had invented logical puzzles which required an investigation of the laws of thought, and that caused a race of so-called teachers of Dialectic to spring up in Greece.

So the first movements in Logic were trammeled by the ignorance and empiricism of those who called themselves teachers.

The experience of our own age has taught us that true science is more impeded and injured in this than in any other way. A whole class of speculative logicians in the early times went by the name of Sophists.

We are accustomed to hear the Sophists spoken of in terms of contempt, and *sophistry* has come to mean *Fallacy*. But we should err greatly, as many in all ages have erred, if we regarded them as wholly evil. The most enlightened writers of modern times have demonstrated that much of the odium which attaches to the name belongs really to the abuse of their art; they were paid teachers—among whom are enumerated Protagoras and Gorgias—whose duty was to train up young men for the duties and pursuits of public life. The character of the Greeks, who were fond of riddles and disputes, and the errors of the age, led to their real *sophistry*, and their abuse of the rhetorical art to make " the worse ap pear the better reason;" after that, their efforts were not for the purpose of widening the range of knowledge and truth, but really served to check these, and thus give a free course to fallacious reasoning.

The Logic of *Euclid* consisted in *negative* proofs; his design was, in encountering an opponent in controversy, not to attack his premises, but his conclusion.

Chief among the early logicians, as he is distinguished among the sages of the world, was *Socrates*.

Much interest and sympathy attach to the virtuous and heroic life and the tragical fate of this wise and good man; but it is principally by his philosophy and logic that he has been useful to the world. Keeping in view always before his numerous scholars the dignity of Logic as a science, and the loftiness of the reasoning powers, he guided the logical processes by what is now called "*common sense.*" "This is implied in Cicero's declaration that Socrates brought philosophy from heaven to earth. Xenophon, likewise, tells us in his 'Memorabilia' that when he wished to form a decision on any subject, his reasonings always proceeded from propositions generally assented to or understood."* Condemning the errors into which the Sophists had been led, he claimed *Truth* as the real aim of reasoning, and established in all his arguments a high principle of moral responsibility. The *analytic* process was that mainly employed by Socrates; and thus, when Plato appeared, he found the science of Logic and the art of Dialectics presented by detached and isolated views as the result of previous investigations. The *analysis* had only prepared for the *synthesis*.

The plan adopted by Plato was the *Synthetic method*, and by this he worked out many great results.

Perhaps the best feature in the Logic of Plato was that, on approaching the science, he tells us to keep the mind free from all preoccupations and preconceptions: he declared, as

* Blakey's Historical Sketch of Logic, p. 24.

an axiom, that "Ignorance is the true start-point for Science." Disputing the assertion of the earlier philosophers that *sensation* was the foundation of truth, he proved it to be one of the instruments by which truth is arrived at. Without stopping to give a sketch of his system, we may state that his Logic and theology are so intimately connected that we may judge of the vigor of the one by the developments of the other. He proved the existence of a Deity who was the measure of all knowledge, the centre of all truth; and in mysterious language he declares that this centre is "the beginning, middle and end of all things." But Plato was to be eclipsed by a greater mind—in fact, one of the greatest minds the world has ever seen.

When much material was thus collected, when many vague theories had thus been started, and when crowds of ignorant pretenders had arisen to be converted or silenced, Aristotle came to create a new system—to enlighten, to harmonize and to sweep away all the errors of the Dialecticians and the Sophists. He who was to correct the characteristic errors of the Greek philosophy was himself a Greek. The Greek mind was eminently a curious one. All the speculations of philosophy, all the systems of Ethics, were directed apparently and nominally indeed to the discovery of truth; but if they reached, by specious arguments, a pleasant conclusion, it mattered little for pure truth. They contented themselves with the fruits of their system, once that system was established.

The Athenians were characterized by the apostle as "spending their time in nothing else" but the pursuit of novelty, and they were but the types and representatives of the other states and cities of Greece. There are in the early Greek authors many corroborations of the apostle's assertion.

Aristotle, building upon the combined foundations of Socrates and Plato, discovered many new principles and established new rules, until he had elaborated the system of

Logic which we have at this day. His logical works, published in full under the title of "Aristotle Organon," comprise the following works: 1. The Book of the Categories; 2. Of Interpretation; 3. The Prior Analytics; 4. The Post Analytics; 5. Topics; 6. Of Sophisms.

Of these the most important are "The Book of the Categories" and both "Analytics." We shall proceed directly to explain their meaning.

He drew the true and somewhat nice distinction between Logic and Rhetoric, and established the fact (a fact not yet learned by many who call themselves logicians) that Logic is not concerned with the truth of propositions, but only with the reasoning upon such propositions as are given into its charge. If the premisses be *true*, then Logic will give a *true* conclusion, but if the premisses be *false*, Logic gives a *false* conclusion; but in this latter case the *Logic* is as good, the argument as valid, as in the former.

In establishing his dictum, which we have assumed to be the universal principle of reasoning, he laid down the general law of Logic—a law which has been misunderstood and misinterpreted, for this dictum was not a model of common arguments, but simply a *test* for all.

As the Greeks looked for truth and found that Logic did not impart it, that before Logic could be used they must be possessed of premisses, which premisses are given them either by *intuition*, by *deduction* or by *observation*—*i. e.*, *induction*—they either abused Logic for not doing what it could not propose to do, or else injured it much more than their abuse could do by using it as a vehicle for false philosophy and mythic religion. They took, to save themselves the trouble of laborious induction in search of premisses, the vagaries of their own quick, joyous and disputatious minds, and thus produced monstrous and absurd conclusions which, since their *Logic* was valid, they felt satisfied to consider as *true*.

The union of this Grecian spirit with the equally vague

and fantastic imagination of the Orientals, with whom by conquest they became acquainted, further corrupted their intellects, and robbed Logic of its true character and mission, leaving the whole domain of Philosophy without the true guide of Reasoning.

Let us now look in turn at the logical works comprising the Organon.

The Categories.

We are in the habit of using the word *category:* for example, we speak of a person or thing being put in this or that category; the word and its use we owe to Aristotle. His categories are *ten* in number. They are not all now considered of importance in classification, but are still worth an explanation as the original system from which, by careful elimination, we have produced our own later classifications. The categories were supposed to imply answers to all possible questions concerning a *term* expressing an act of apprehension—*i. e.*, all of which we can have any knowledge.

1st, Substance; 2d, Quantity; 3d, Quality; 4th, Relation; 5th, Action; 6th, Passion; 7th, The Where; 8th, The When; 9th, Position, in space; 10th, Possession.

The categories may be thus more fully explained:

1. SUBSTANCE may be defined that which is in itself, which may be conceived as existing by itself. This is divided into *spiritual* and *corporeal*, and subdivided according to *classes, genera, species*, etc.

2. QUANTITY may be translated *how* much or *how great*, and by implication, *as to time, how long*. Thus, under the head of Quantity, we have the three special considerations of *Number, Magnitude* and *Time* (as to duration). *Number*, we know, is either abstract or concrete, as when we speak of a number disconnected with any objects, or of a number of objects and things. Thus, *quantity*, as a category, covers the science of *arithmetic. Magnitude* is either *linear, superficial*

or *solid;* and thus its genus *quantity* covers, likewise, the science of *geometry*. *Time* is either permanent or successive, and is used to indicate the movements or conjunctions of *Number* and *Magnitude*.

3. QUALITY describes the kind or sort of which a thing is, and is subdivided into *Habit*, or a quality induced by frequent repetition of the same act, as *virtue, vice*, etc.; *Inherent nature*, as man's *reason*. From these grow the many subdivisions of color, sound, hardness and shape.

4. RELATION is the consideration of two or more objects with reference to each other. The first object *of two* is called the relative, the second the correlative, as *prince* and *subject, master* and *servant*.

5. ACTION has a double meaning; it is at once the exertion of power by one body on another and the effect produced by such an exertion.

6. PASSION is the endurance of another's action.

7. THE WHERE includes the three meanings which we express by the words *where, whence* and *whither*, as *in Philadelphia, from New York, to London*.

8. THE WHEN has reference to the *exact period* of *time*, and not *its duration*, which, as we have seen, belongs more properly to quantity. *The When* may be expressed by the phrases *to-day, to-morrow, a hundred years ago*.

9. POSITION has reference, not to the *place where*, but *to the posture in which*, a body is found, as *lying down, standing up, kneeling*, etc. The question then is, *how* did you find it? not *where?*

10. POSSESSION has reference to something belonging to the object, or placed upon and clothing it, and as a category covers all questions concerning the rights of property.

Of these categories it will appear that *substance* stands apart from the rest in that it is sensibly existent and they are all *attributes* of such an existence. It will further appear, upon examination, that *Quantity* and *Quality* are *essential* attributes, *i. e.*, belong to the essence of the object *necessarily;* while *Relation, Action, Passion, The Where, The When, Position* and *Possession,* are accidental circumstances which may be dissociated from it.

To render this clearer for facility of reference, we state it in a tabular form. In this table we place all the explanatory parts as by the rules of division before given, but *number* the categories that the eye may at once rest upon them.

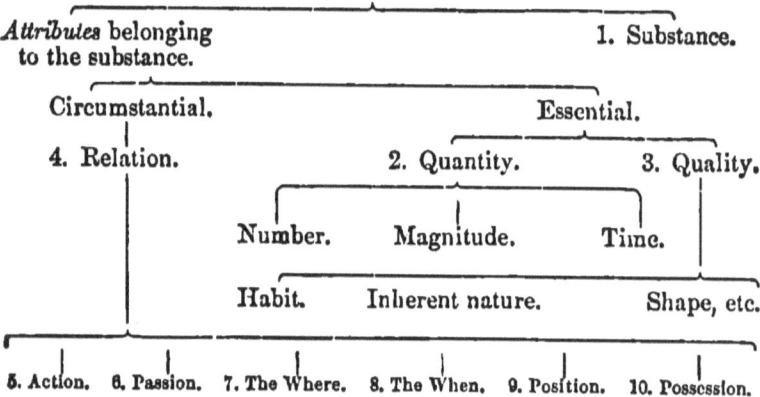

Aristotle asserted that everything which could be said of any subject is included in one, some or all of these categories, and his own illustration of their use is one of the simplest which can be found. It was as follows: "Substance, *man;* Quantity, *one;* Quality, *white;* Relation, *greater;* The Where, *in the Forum;* The When, *yesterday;* Position, *sitting;* Action, *whatever he may be doing;* Passion, *whatever may be being done to him.*"

It is under this first attempt at method that the sciences began to range themselves in classes, and by this all other systems of classification seem to have been suggested. Thus, *Substance* is the foundation of all Physical and Historical investigation; Quantity, the subject of Mathematics; *Quality,* of Medicine; Relation, of Ethics; Action and Quantity, of

Astronomy, Music and Mechanics; *Passion* and *Action*, of Electricity; *the Where*, of Geography; the When, of Chronology; *Position* and *Quality*, of Sculpture; *Habit* and *Position*, of Painting; and so each art and science would be found to range under one of these singly, or more than one when combined.

The books of "Prior and Post Analytics" originate and develop his system of the doctrines and use of the Syllogism. They have been the resort of all writers on formal Logic since his time, and there has been but little alteration in his method. Aristotle established but three figures of the syllogism, the fourth being afterward added by Galen.

In his book of Topics he discusses the subject of *Predicables*, or *Classes*, and establishes the expression of a predicable to be in four ways; *i. e.*, by *genus, differentia, property* and *accident*; in these he implies the *species*, since we have seen that if we add the *differentia* to the *genus* we obtain the species.

In his book of Sophisms he states thirteen Fallacies as including all those which can bear a syllogistic form. Six of these refer to the *words* used, and are called *Fallacies in dictione*, and seven consist in the *matter* of the propositions, and are called *Fallacies extra dictionem*.

The logical works of Aristotle seem to have been providentially preserved. Transmitted by his disciples from hand to hand, they were at length concealed in a vault during one hundred and thirty years, until they had mouldered into an almost illegible condition. Restored from this condition, they came by the fortune of war into the hands of a Roman general, and thus were given a second time to the world.

We cannot pause to notice all the changes attempted in Logic and Philosophy from this time until the Christian era. After the Peripatetics came Pyrrho of Elis and his *Skeptics*, who seem to have employed Logic to deny the possible attainment of pure truth. They embodied their system in *Ten Tropes*, or logical rules for the government of mind in the search of truth. Their doubt led to what they termed a *suspension of judgment* rather than a positive denial.

Of the Epicureans and Stoics, it may be said that they aimed at the establishment of no Logical system, but rather a few tenets in the shape of propositions; by these, as doctrines, they guided their course.

The tenets of Epicurus may be comprised in the assertion that "whatever is useful, pleasant and delightful is *true.*" This is to assert that man's senses and bodily appetites are the only test of truth. These have been called his "emotional criteria."

The Stoics rejected the categories of Aristotle and adopted four of their own, and attained the conclusion that "pain is no evil"—a philosophic stretch of the imagination which has given its name to an unshrinking endurance of pain and evil.

Very little transpires concerning *Roman systems* of Logic. Although Cicero, Maximus of Tyre and Galen lay claim to the title of logicians, the logical system of Aristotle was adopted by them all; Rhetoric became the more valued and important study.

The history of Logic, then, from the time of Aristotle to the coming of Christ, is not a history of change, but the logic of Aristotle, however unchanged, had been most unworthily used. No longer the guide and test of just reasoning, it became the vehicle of ingenious falsehood, was made to support any theory and gave power to its possessor "to argue on both sides of any question." To satisfy curiosity it established any paradox, and one being made the premiss to another, the error was multiplied "in infinite progression undefined." It was not the logical system, but the mind of man, which needed purification—not abstract propositions, but the matter they contained, which demanded scrutiny.

We shall see also that the misconception of the sphere of Logic was equally fruitful of error long after the establishment of Christianity, and that it has remained for the nineteenth century, notwithstanding the utmost resistance of many learned but dogmatic philosophers, to give to Aristotle and his system their true place in the domain of science—an *instauration* not by one man, a new Organon not the product of one teeming brain, but the tribute of Philosophy, inductive and deductive, to Aristotle, the great founder and framer of that system which alone controls the unbridled reason and sends pure truth into the channels of usefulness and practice.

But, meanwhile, the coming of Christianity was to produce great marvels in the domains both of Logic and Philosophy.

(57.) The Logic of Christianity.

The Logic of the Grecian schools had been the guide of man's Reason, but now it was itself to be brought into companionship with a higher human attribute, Faith. Premisses were no longer to be sought by the ordinary means of evidence, but to be supplied in a new and marvelous manner. Christianity combined this new element with Philosophy, and taking the art of Logic as the vehicle of its great truths, used it in a manner at once beneficial and practical, putting an end, as it seemed, to the controversies and paradoxes which had beguiled and engaged the Greek and Roman mind.

By this new tutelage of human reason, Christianity produced an immediate and startling change in Philosophy by opening the *Finite* upon which man may use his reason, as well as indicating the *Mysterious* and *Infinite* to his faith.

As much as we may despise the Greek systems of speculative Ethics upon which they employed their nobler Logic, we must remember that they were the gropings of men in the dark, pursuing a faint glimmer of light in the hope that it would lead them into the full sunshine and free air of Truth. They had no revelation of intelligible fact or of mystery. The efforts of Plato to attain to different degrees of knowledge which he calls "the absolute, the probable, the imperfect," the Politics and Ethics of Aristotle, the bold dicta and quiet endurance of the Stoics, the "emotional criteria of Truth," propounded by Epicurus, and so much abused by his disciples, were all vain attempts to arrive at that knowledge which could come to man only by miraculous revelation. God vouchsafed no such revelation to them; it is no cause of wonder that they erred greatly without it.

This, then, was the crowning glory of Christianity, that it gave to man pure Truth, and furnished him with a world of new facts upon which to reason, of glorious propositions upon which to try the powers of his Logic. They furnished him a boundless field, with the word of God as a beacon infallible, and where reason could not obtain internal or analytic evidence, resting its judgment on external evidence as a basis. God said to man, Believe and ye shall be saved.

Unlike the Greeks, the Jews had always possessed this revelation in a ceremonial and progressive form. Their own Scriptures had disclosed to them not only the true story of man's origin and fall, but of God's supremacy and his gracious design of restoration, and their prophets had told them with a heavenly Logic of Type and Symbol, premiss upon premiss in glorious abundance, of that certain conclusion, the advent of the Messiah.

The "fullness of time" came, and the event fulfilled the prophecies, the conclusion completed the premisses. Christianity brought philosophic as well as religious light.

By a strange infatuation, they who had thus awaited His coming refused Him when He came; and since He could not be the glory of His earthly "people Israel," He was, in a truly philosophic sense, "a light to lighten the Gentiles."

In three centuries, He had been eagerly embraced by heathen Rome, and the Logic of Aristotle, freed from its vile and improper uses and used as the propounder of a full and pure creed, was applied with great power to the spread of the Christian religion. Where false premisses had been ignorantly used, leading to a false conclusion, or where false conclusions had been improperly deduced from true premisses, everything for a time was changed. Truth was everywhere triumphant, and its reign seemed to be eternal.

Such was the first influence of Christianity upon Logic. Containing in itself nothing repugnant to reason, it gave a host of new and glorious truths fresh from the mouth of God; it simply threw away the vague speculations, the unsound paradoxes, which had been heretofore used as premisses, and took these new *truths* to reason upon. In the teachings of our Saviour and the apostles, it need scarcely be remarked, not only that every statement is true, but that every argument is valid.

On the other hand, Logic, turning gladly away from the subtleties and absurdities of mythical philosophy, pressed forward with ardor in the task of systematizing and promulgating the new doctrines of Christianity.

In this manner arose the logical systems of the early Christian writers and apologists known as "the fathers." There is, indeed, error to be found in their uninspired writings, such as we should expect in all human productions, but, from Justin Martyr to St. Augustine, one object of their writings seems to have been the harmonizing of Christian doctrine with the Logic of Aristotle, and thus, while they preached the truth, to show at once the union and true relation of Reason and Faith. How well they succeeded as a class may be seen at the present day from the growing interest in their writings which is manifested by all who are interested in Religion or Philosophy. Never forgetting that they were surrounded by enemies and error, one part of their works was fiercely controversial, always keeping in view the *elenchus*, and warily observing an opponent, or rather the many opponents who were scrutinizing their deeds and words.

Where, in the old system of Philosophy, *Sensation* was the

starting-point, and man must evolve philosophy from within himself, they established *Revelation* as the centre and starting-point, and would draw, by the same logical formulæ, all true philosophy from God. From this time Logic was inseparably connected with theology; the Church ruled the world.

The Christian Church had, in its union with the Roman empire, a strength and stability from which great philosophic results must have sprung; but just when they were framing this glorious system at once of Religion and Philosophy, the Roman empire of the West fell under the ruthless attacks of the Northern barbarians, and the Church was temporarily paralyzed by the shock. For centuries after, the great efforts of the Church were directed to the attainment of a firm social basis and political power.

We have already stated the connection between Logic and Philosophy. They may be dissociated, but are both then useless. Thus, indirectly, Philosophy has exerted such an influence upon the uses of Logic that it is important to trace the systems with which Logic was combined, and to promulgate which it was used after the establishment of Christianity. Most of the Christian writers investigated the subject of the human reason, and studied the Logic of Aristotle.

As might be expected, so magical a transformer as Christianity was not without fierce philosophic opposition. With equal step Skepticism and Heresy advanced. Those who were doubters before where only *Science* was concerned were doubly doubters when told of Christian *mysteries*.

The representative of the new skeptics was Sextus Empiricus, who lived in the beginning of the third century, and who was but a new incarnation of Pyrrho of Elis. Unwilling to receive, on *prima facie* evidence, the truth of the new revelation, they had fallen back upon the old material, and had worked to the same results as the Greek philosophers; they turned their backs on the light—which admits of no better proof than the physical light of day—and walked into the cave of darkness, of doubt and, in a religious view, of despair.

The skepticism of Pyrrho, three hundred years before Christ, was consistent and well deduced when compared with this, and yet the Greek academicians, we know, had convicted him of absurdity. "Because everything is contradictory, everything is false." Now, if *this* be *true*, the axiom itself is false, and so the skeptic, thrown upon the horns of a dilemma, must grope again in vain for new proofs of falsehood and new certainties of doubt.

Of the Neo-Platonic, Eclectic or Alexandrian school, the object seems to have been to unite the Greek philosophy and Oriental dogmatism into one system, but it was a false and feeble combination, fated to a speedy and ridiculous end.

Its metaphysics, as prepared by Plotinus, was the attempt by the combination of heathen obscurities to attain to Christian light; its theology, as reduced by Iamblichus, was a strange retrogradation from the Scriptures, which revealed the person and word of God, to the ridiculous deities of the Pantheon; and its Logic, of which the great Porphyry was the applier, was an attempt, by the use of the Aristotelian system, to establish all these errors, at the expense of the fair fame and even of the existence of Logic.

Nor in the singular application of Christianity to Logic must the Gnostics be forgotten. Their name indicated their creed; γνωσις, *knowledge,* as opposed to *faith:* naked Logic, stripped of its armor, was made again to do duty in the ranks of the Prince of Darkness. Gnosticism "took such portions of the Gospel as suited its views or struck its fancy, but these rays of light they mingled with such a chaos of absurdity that the apostles would hardly have recognized their own doctrines."*

The greatest perhaps of the indirect evidences of the truth of the Christian religion is that, in spite of the false systems

* Burton's "Heresies of the Apostolic Age," p. 15, quoted by Neil.

which sprang up to oppose it, it has steadily and mightily prevailed; in its progress it has purified human philosophy and unfettered Logic; but it did not accomplish this without fierce contests; it was to come upon dark days in which it was the only glimmer of light—days in which the misuses of Logic were no longer to be confined to profane systems or heretical creeds.

Then came the Schoolmen or the so-called Scholastics.

The *first* era of Scholasticism was the adoption of Logic as the form and vehicle for Religion, and thus far they were in the right path.

The *second* phase was the attempt to unite Religion and Philosophy, and this produced new champions of *Realism.*

The *third* phase was an opposition; Religion and Philosophy were rudely dissevered, and this produced *Nominalism.*

If, now, we separately consider these three phases of the Scholastic philosophy, we shall perceive that the first was the just and true one, and that the succeeding ones were learning which had to be unlearned.

That part of the Greek system which could be made the *form* and *vehicle* of religion, as it is of all correct reasoning, was only the Logic. To apply that to the service of Faith was just the first design of Christianity toward Logic, and thus far the Schoolmen were right—indeed, it would seem ignorantly right, for while using the forms which constitute Logic, they still persisted in calling many other parts of the Greek philosophy by the name of Logic, and thus making Logic bear the blame which truly belonged to the errors, obscurities and absurdities of exploded systems of metaphysics, theology and morals.

This is apparent in the works of Alcuin, the contemporary and friend of Charlemagne, and especially in his dialogues on "Grammar, Rhetoric and Logic."

Lofty was the simple distinction of St. Anselm that there are but two modes of Cognition, Faith and Science, and grander yet the idea, "that Science begins where Faith ends"—in the bosom of God!

But let us consider the second and third phases.

Nominalism and Realism were but the reproduction in the ninth century of the old Platonian controversy already referred to. *Nominal* and *real* were the abstractions of what we call respectively *universal* and *particular*.

When I speak of a single man, and point him out, I designate a real, existent individual; when I speak of *man* as a common term, is there a real entity corresponding to the word? The Realists said, Yes! the Nominalists said, No! it is but a name to indicate numbers. This had been the origin of the controversy.

Plato, with his divine but vague philosophy, had asserted that there was a *real* existence, an archetype in the bosom of God corresponding to the name of a class, as *man, angel;* Aristotle, that they were only generalized names from many individual abstractions. And thus these great parents of Logical Philosophy set the example of wrangling to their myriad children of the schools. It is curious to see how such a dispute first connected itself with religion. It was thus: the question seemed to involve another and a more important one, viz.: "What is the foundation of human knowledge?" Roscellinus of Compeigne, who lived in the eleventh century, was the originator of the new controversy in the Middle Ages between the Realists and the Nominalists. He was a fierce *Nominalist;* and as this led to supposed heresies, he was an object of persecution on this account. As warmly was the cause of *realism* espoused by William of Champeaux, and throughout the schools there was a word-war of great fierceness on this subject.

Passing over the quarrels of the Schoolmen until we reach the time of Roger Bacon, and thus neglecting many great names in the history of Logical Philosophy, we are struck with the power of his experiments and analyses, and the manifest fact that he deserves the name of the founder of Experimental Philosophy—that his "*Opus Majus*" may justly be considered the precursor of the "*Novum Organum*" of his more illustrious namesake, Francis Bacon.

Disgusted with the categories of Aristotle as trammeling an ardent physical scholar who must establish categories for himself by experience, he considers *experiment*, based upon constant *observation*, the only rule for philosophy, and in his works in the laboratory and with his pen we discern the first dawning of the day of Induction.

For a while, as was very natural, formal Logic fell into disrepute, and gave way to experiment in physics; and from that day down to our own times, there has been but little appreciation or understanding of the art of reasoning, although it has been constantly used and constantly ignored. Like savages who breathe the invisible air round them and are not aware of its existence, so minds of all kinds and calibres have used the Logic which they found established as the vehicle of thought without knowing where to make their acknowledgments.

(58.) The Logic of Experimental Philosophy.

Now an element seems to have been introduced into philosophy which till then had been considered unimportant, and that was *observation* and *experiment;* or, to use the term by which we have expressed the methodical and successive observations of such phenomena in nature as will lead us to general laws, Induction. Aristotle himself had stated the value of induction for the discovery of new truth, and men, in all ages, had used it as an exercise of *common sense* in their ordinary conduct; so that it must not be supposed that, in any sense, Bacon is its inventor. He only applied it by system to natural science.

Logic, which is the vehicle of truth in its intellectual passage from premiss to conclusion, had only reasoned upon the *known* and *conceded*—mainly from some *general* law to a *particular* example; now its premisses were to be new truths aggregated by experiment; it was to reason from many particular examples to the establishment of a general law.

Bacon was the early interpreter of Nature, Descartes more especially the analyzer of Thought. To each is due an illustrious share of the developments in philosophy. But Bacon is the more distinguished because his investigations were made in every domain of nature, and his system is at once more intelligible and popular on that account.

The starting-point of Bacon's philosophy was the assertion that the *universe is a great storehouse of facts*, and that it is man's duty and interest, and it ought to be his pleasure, to explore, discover and understand these facts, not only in their isolated characters, but in their relations to each other and to the universe itself. His experiments and his use of the experiments of others were to enable him to arrive at general laws of the universe. Now, corresponding with the world around us—that is, the world of nature—there is a world within us, the world of Thought. Let either be impaired or cease to exist, and in just such a proportion is the other impaired or does it cease to exist.

To unite them we have sensation and perception, and the union is lost if sensation and perception fail.

The happy union, then, of Thought and Nature, would lead man to Truth, and to attain to Truth is his highest aim. It will at once be seen that this was the establishment, not of a logical, but of a philosophical system. But to proceed: the various forms which truth assumes in order to inspire the faculties and entice the pursuits of men are called sciences, and by an examination of multitudes of these phenomenal facts the true definitions of the sciences might be made, their true relation determined and a plan of classification formed for practical purposes.

Such, then, very briefly, was the aim of the new experimental philosophy—a great restoration which was proposed by Bacon in his *Instauratio Magna*. With it directly Logic had but little to do, but that little led men of science into errors which remain to the present day.

Without attempting to enter into the details of the "Great Restoration," it will be well to consider some of the steps proposed by Bacon as preliminary to it. Finding, in his inquiries about facts or *phenomena*, that they greatly differ in importance—that some are simple, others complex, some are easy of interpretation, others very difficult—he proposed a classification of the *instances* in which any *phenomenon* or fact occurred, and this should be a sort of value scale of the instances in which a special phenomenon occurred. These he calls *prerogative instances*, or those cases of most importance to us in interpreting a fact or a series of facts. He has stated *twenty-seven* of these, from which we shall choose *six* as better illustrating their own meaning than it can be done in other words. Our purpose is not to use these, but merely to indicate their nature and design.

I. *Solitary instances*, or those in which two or more objects *agree* or *differ* in all qualities save *one*. Thus a rabbit-skin and a piece of rough glass, which differ in all other qualities, agree in this, that on being excited by a metal they both become charged with positive electricity, while two pieces of silk ribbon, only differing in color, when thus excited, become the one positively and the other negatively electrified.

II *Forth-showing instances*. Under this head range those *facts* or *instruments* which show forth the quality in question in the highest degree, as a galvanic battery in electricity and a barometer in pneumatics.

III. *Analogous instances*. Those in which are found objects bearing a resemblance of purpose or relation, however unlike the objects themselves may be. Thus, a camera-obscura is analogous to the eye and a system of watermarks to the heart.

IV. *Crucial instances.* There are two probable meanings to the word *crucial* as here used. It may be the putting nature to the torture, the crucifying her, to wring from her her secrets, or it may have reference to the wayside crosses which at the parting of the roads indicate the true direction to the traveler. Franklin's electric kite might be called a *crucial instance,* in the first sense. Such also, in the second, was Newton's law of gravitation, a finger-board for ever to point to the true direction of investigation and belief concerning our solar system.

V. *Varying instances (Instantiæ migrantes).* Those propensities of bodies which change to a greater or less degree. Among these would be included change of form from solid to liquid and from liquid to gaseous, and the reverse.

VI. *Companion and hostile instances.* Of the first would be qualities which usually accompany each other, as heat and flame; of the second, those which are never in conjunction or alliance, but seem to repel each other, as the positive and negative poles in electricity.

The other instances, which we cannot stop to mention, are designed to exhaust the classification of experiments on facts, and to lead to induction; and here began the danger and difficulty; it was here, also, that the syllogism which Bacon despised and misunderstood was and always is the only safe guide of Philosophy. For, suppose the facts ranging under these instances to be established, how many of them will give us the right to the establishment of a general law or a distinct science? We have seen that, in most sciences, we only attain to *likelihood.* On account of human ignorance, the process has been this: we first observe a few facts; we then adopt a hypothesis based upon them—*i. e.,* jump at the general law—simply in order to make a *nidus* for our accumulating facts, and thus proceed to verify—if the new facts will verify—our proposed theory. The tendency of man's mind is so great, however, to repose upon a darling hypothesis, even if it be unsound, and rather to seek, like an advocate, for such facts and statements as will support it, than to look for just proof, and in the absence of such to discard it, that induction has often led to grievous error. Many a student has learned on hypothesis some part of Natural Science, and when he had just mastered it has been obliged to discard it for another.

In the consideration of *Judgment*, Bacon has given special attention to the fallacies which assail the mind of man. These he calls *idols of the intellect*, and in almost every case, since they are contained in false judgments, they belong to the class of material fallacies. But all these idols occasionally assume the garb of logical fallacies.

These idols, or εἴδωλα, which Bacon calls "the deepest fallacies of the human mind," are the sources of error which assail men in their investigations in Philosophy, and which "must be renounced, and the intellect wholly freed and purified therefrom," before we can hope for healthful progress. By the word *idol* Bacon means the prejudice which stands in our way of receiving truth and the bias of the mind from which such prejudices arise.

But these idola will most clearly explain themselves; they are of four classes—*Idola Tribus, Idola Specus, Idola Fori, Idola Theatri;* and with reference to these an author of his own time remarks: "The temple which he purified was not that of nature itself, but the temple of the Mind; in its innermost sanctuary were all the idols which he overthrew."

1. The *idols of the tribe* are those which are imposed upon the understanding by the general nature of mankind; in other words, they belong to the *human* tribe, in its universal comprehension. Thus, he asserts that men, as men, are quicker to be moved by *affirmative* and *active* events than by *negative* and *privative*, though in justice they should be moved by both. To illustrate this, he tells the story of the Greek who was shown in Neptune's temple the votive pictures of

those who had escaped shipwreck; and when asked if he did not now acknowledge his divinity, said: "Show me first where those are painted who paid their vows and were then shipwrecked."

2. The *idols of the den or cave* spring from the nature of each particular man, and grow out of his peculiar features both of mind and body; these may also be fostered or developed by education, custom or accident. The name is suggested by fancying the confusion and error of a man being brought out of a dark den or cave into the full light and glory of nature. This finds its counterpart in the world of philosophy, where men only emerge from the den of their minds to find confusion and disorder in the beautiful universe of God.

3. The *idols of the market* are errors which grow out of *words* and *communication*, such as are the pass-words and common coin of conversation and intercourse in the market-place; and they imply, like the idols of the tribe, a social organization, but on a much more limited scale. Instead of being universal with men, they are errors which belong to a small circle, like a crowd in a market-place, moved, at the sound of an orator's words, by a common impulsion of prejudice, passion or other emotion. These idols are causes of the greatest disturbance, as they are immediately connected with the naming of things, "for words are generally given according to *vulgar* conception, and divide things by such differences as the common people are capable of; but when a more acute understanding or a more careful observation would distinguish things, better words murmur against it."

Thus, many words in our every-day use convey no definite meaning to the mind, but have, in their very indefiniteness, so many shades of meaning that they are a constant cause of verbal fallacy. As special reference has been made to such words in the chapter on Fallacies (X.), it will only be necessary to mention a few such to illustrate the idols of the market-place; such is the word *republic,* which we have been apt to confound with *democracy; Liberty* means either *freedom* or *license,* as its champions wish, and *taste* and *beauty* have as many forms as there are eyes to see or imaginations to indulge.

The last of the sources of error enumerated among the idols of Bacon are the *idols of the theatre.* These he distinguishes from the others as perhaps of more social power and influence. Of these he says, "They are superinduced by false theories or philosophies, and the perverted laws of demonstration." They are comprehended under three heads, *Partisanship, Fashion* and *Authority.*

Partisanship is the generic name under which are found factions in politics and in religion, and under whose influence wars of creed and caste have so often desolated the world.

Fashion is a kind of partisanship which, however, has few opponents and no great rivalries, but which pervades society from high to low. We do not refer to its simple sway in dress, equipage and social life, but to its more comprehensive dominion over all the works and thoughts of man, over art, science, religion. Great masses of men are herded like cattle and driven willingly in the train of this all-swaying Fashion, resting their happiness here and their hopes in an eternal future upon the dictum of Fashion.

As Fashion partakes of the nature of Partisanship, so is *Authority* strengthened by an alliance with both. This consists in blind obedience to an existing control and reliance upon it without the use of our own judgment.

As God, who has given man Reason, has made some things higher than that reason, but nothing repugnant to it, every theory of authority in Church, in State or in general philosophy is, of right, to be examined by our reason before we can accord to it our belief. Reliance upon authority, without a due understanding of its claims, is to treat our own moral constitution with injustice, and to stop the wheels of healthful progress both of individuals and societies.

In reviewing these error-sources it is scarcely necessary to remark that it is the abuse and not the use of our words and associations which lead to them.

Thus, the *idols of the tribe* would not be false and deceitful if man should concur universally and everywhere in just and truthful opinions, nor would the *den* darken men's minds to the true light if they were capable of carrying into their meditation the true elements of combination and just views of the objects in the universe around them. Heraclitus has told us "that men seek the sciences in their own narrow worlds and not in the wide one." Such is the influence, but not the necessary consequence, of the den.

So it is easy to avoid the errors which grow out of ambiguous words, such as those which mark the idols of the *market*, by demanding just definitions, and when such cannot be given either agreeing *for argument's sake* upon one which is not just, *or* declining to argue at all where the very question is involved in obscurity.

We may observe, concerning the *idols of the theatre*, that partisanship has its good as well as its evil character, and that to championize the right is noble and just; it is, however, even in such a cause that its tendency is to extremes.

So *fashion*, crowds of whose votaries are miserable and self-tortured, is incident to man's social character, and is productive to those who use it aright of method and comfort and success. Although fashion has done much evil, it could not be spared in our social or intellectual systems. Nor must *Authority*, however formidable the name, be accounted of slight importance, for under just authority are ranged *obedience, order and wholesome discipline;* without it government would be anarchy, and education would be a curse instead of a blessing. It is the time-honored *abuse* of it which demands our dislike and resistance.

Beyond a few and very erroneous allusions to the Logic

of Aristotle, Bacon and his immediate successors did very little for it as a science.

Hobbes seems to have just views of the syllogism, as "the instrument of demonstration," but carried his investigations, his written ones at least, very little beyond such a statement.

Resting upon the basis of the Baconian philosophy, the thinkers of the seventeenth and eighteenth centuries seem to have neglected the *art of reasoning* for the *subject-matter* about which we reason, and thus to have entirely confounded Logic with the art of thinking. For this they had the authority of their great master, Bacon, who, in his "Advancement of Learning," has divided the Art of Judgment into *Induction* and *the Syllogism*; and has classified as four kinds of demonstration: 1. That by immediate consent and common notions; 2. By Induction; 3. By Syllogism; and 4. By Congruity. The error of this classification is at once apparent to us.

Indeed, it may justly be said that, in everything pertaining to *Logic* in its proper meaning, Lord Bacon is entirely at fault, while in everything which bears upon Experimental Philosophy he is great beyond any competitors, for he is its founder; and as a few words have shown that all induction must be brought to the syllogism to verify and test the laws at which we arrive, his philosophy can be easily disconnected from his Logic, and the faults of the latter exert no evil influence over the excellences of the former.

Many logicians in England, France and Germany followed in the steps of Bacon in the seventeenth century, attempting to unite Logic and Experimental Philosophy in a manner which was injurious to the former.

Locke, misunderstanding the syllogism, as Lord Bacon had done, discards it from his system, and bases his views of the understanding on *two sources* by which ideas enter the mind, viz., Sensation and Reflection. But to show how so great a thinker rebukes himself, he states reasoning to consist of *four* parts: 1st. Finding proofs; 2d. Arranging them; 3d. Showing their connection; and 4th. Employing them correctly.

Now, what is all this but, 1st. Finding middle terms by which to establish premises; 2d. Stating syllogisms; and 4th. Combining arguments? As for the 3d, that is included in the 2d, for they cannot be arranged without their connection being manifest.

Leibnitz, in Germany, seems to have thrown light upon the theories of Descartes, and to have elucidated also many things in Locke.

Milton has been called the most learned man of his age; he vindicated this opinion by writing upon almost every subject within the range of knowledge, and in most cases writing well. We are not, therefore, astonished to find that he has written a work on Logic. It is in Latin, and seems to be very little known. In that he adheres to much of the Aristotelian doctrine, and specially championizes Peter Ramus, the logical Martyr. He divides Logic, which he calls the chief of Arts, into two kinds—*Natural, i. e.*, the faculty of reason in the human mind; and *Artificial, i. e.*, rules for directing the operations of that faculty. But even Milton erred in stating that "it belongs to Logic to lead us from universals to particulars," which would limit the Syllogism to Deductive reasoning.

In this state of confusion Logic existed until the new rise of Philosophy in the eighteenth century, the source of which was the continent of Europe rather than England.

(59.) **Logic in the Eighteenth and Nineteenth Centuries.**

But little remains to be said in order to complete this brief sketch of the History of Logic. Even to mention the names of the principal writers who have sprung up under the impulse of the Baconian philosophy from that time to the present would occupy more space than we can give, and to discuss their metaphysical works would in this connection be difficult and improbable.

The logicians of the eighteenth century seem to have bent their energies to the task of classifying the science, of making such a logical arrangement as would make much labor unnecessary and find for each its true niche in the temple of Truth.

In England, Dr. Isaac Watts published a treatise on "Logic, or Right Use of the Reason," which is a compound of Logic and Philosophy alike injurious to both. Selecting a few tenets from Aristotle, from Lord Bacon and from the Schoolmen, he has endeavored to harmonize them. In another of his volumes, "The Improvement of the Mind," he has moved upon surer ground and with much better success.

Bishop Berkeley wrote the "Principles of Human Knowledge"—a work of profound thought and excellent reasoning; and Bishop Butler has exemplified the correct use and application of Logic in his famous treatise on the "Analogy of Religion."

France has also produced in the eighteenth century many fine logical minds who have devoted themselves to science specially in attempts at classification; among these were D'Alembert, Diderot and their coadjutors, known as the Encyclopædists, who, in the eighteenth century, startled the world not less by their methodical arrangement of the sciences than by the skepticism which their studies induced, and the atheism or denial of God's existence which took the place of doubt.

It would be impossible in a treatise of this kind to do more than simply refer to the present writers on Logic and the present condition of the science.

Archbishop Whately has renewed the Logic of Aristotle in its pristine vigor and placed it in its true position as the only sure guide or Art of Reasoning. Many English writers have differed from him, some in his conception of the mean ing and scope of Logic itself, and others as to the extent to which the Aristotelian system may be carried.

Of the first may be mentioned Mr. J. S. Mill, whose work, according to the view we have taken, may fitlier be called "an encyclopædia of philosophic tenets connected with, or resulting from, the Science of Logic."*

Of the second are Sir William Hamilton and Mr. Augustus de Morgan, who would develop more than *four* categorical propositions and establish what we have called the "New Analytic," and yet they differ from each other in their establishment. Hamilton, the most distinguished philosopher of his age, has numerous followers, among whom are Thomson, who has reproduced the Hamiltonian Logic, in an abridged form, in a small volume called the Laws of Thought.

The most important changes, however, in the applications of Logic to science, are to be found, as has been said, in the subject of Categories and Classification, and to this, in illustration of the later movements of the science, we shall now give a few words. It will be at once perceived that the object is to reach a *summum* genus under which all the sciences may range, and then by a logical *tree of division* to place all the lower classes and their co-ordinate species in their proper places. In any less general classification it is evident that the principle of classification will be changed for the different sciences.

(60.) **Of Categories and Classification.**

This is a part of the duty of Method.

The Categories of Aristotle, which have already been explained, may be considered the basis of the classification of the sciences; for although there has been, in former times, much dispute concerning their true reference—that is, whether it be to words or things or conceptions—it is now allowed that, imperfect as they are, they are designed to apply to the *summa*

* Neil's Art of Reasoning, p. 234.

genera under which all things which are named may range themselves. This establishment of proper *summa genera*, then, is the true start-point of classification.

Many writers have simplified these categories mainly by reducing the number. The schools of Pythagoras, Plato and Epictetus had each its corresponding list or table; Locke wrote three, viz.: *Physica, Practica* and *Semeiotica*, or, as they have been translated, *Substance, Modes* and *Relations;* Hume, two, viz.: *Ideas* and *Impressions*.

Among German philosophers and logicians, Kant holds the highest place. His views are principally set forth in his *Critique of Pure Reason*. He established as an instrument for a pure science of nature the following categories, logical and transcendental:

		Logical.	*Transcendental.*
I. Quantity.	{	Universal.	Unity.
		Particular.	Plurality.
		Singular.	Totality.
II. Quality.	{	Affirmation.	Reality.
		Negative.	Negation.
		Indefinite.	Limitation.
III. Relation.	{	Categorical.	Substance.
		Hypothetical.	Cause.
		Disjunctive.	Reciprocity.
IV. Modality.	{	Problematical.	Possibility.
		Assertory.	Necessity.
		Apodictic.	Existence.

Under these twelve categories all forms of our sensible experience may be brought. This was only part of a system of philosophy, including, besides Logic, æsthetics and metaphysics.

But these are manifestly none of them of that practical form and character which is desirable for useful reference, and hence it has been the aim of later writers, especially upon Metaphysics and Logic, to write out tables of classification which should comprise and methodize all forms of human science. To classify palpable, tangible objects is to arrange them in groups according to a certain method, and that method will usually be based first upon the great division of kingdoms, and afterward upon the relation of species to genus.

If we reflect for a moment upon the innumerable forms of life and existence in the three great kingdoms, Animal, Vegetable and Mineral, we shall at once be struck with the difficulty and labor of a just and adequate classification; and yet, strange as it may seem, true progress in any of these branches has but kept pace with such a classification, the naming and placing of a minute species in its proper place being the necessary way of fixing it there for ever.

It has already been said that the basis of physical classification is the establishment of the *summum genus*, and that the rules of logical division must determine all the *subaltern genera* and *species*. This must serve us for the classification of the known and determined, but in the world of *Theory* another mode may with propriety be adopted: it is the classification by *series*, investigated by Comte. It consists in selecting some particular phenomenon the laws of which are to be investigated, and then ranging the various objects which sustain a relation to it in a nearness proportional to that relation.

With this subject of classification *scientific nomenclature* is immediately connected, and it will appear how important this must be regarded when we consider that the value of the classification will depend upon the names of the different classes, as to their *precision* or *total want of ambiguity*, their *completeness* or *expressing the whole of the class specified*, and their *expressiveness* in *denoting the properties of the object* and the *reason of its classification*. Thus, in Chemistry, a law of nomenclature has been formed, based, indeed, upon some unfortunate beginnings which have been allowed to remain but very systematic and universal in its reception.

But the high aim of metaphysical philosophers to smooth the paths of Logic has been, not the classification of one science, but the analysis and classification of universal Science, the establishment of a complete table in which all human investigation should find its place and link itself to the great mind of all ages in its study of all topics within its sensual or intellectual range.

It will not be attempted to give a history of classification, nor to prepare or copy a complete table of any previous author, but rather to indicate the manner in which it has been done, with a general reflection upon the results attained. Classification, to be logical and just, must be made after certain investigations which are necessary to determine the true class of the object in question. This will be done in Physics by formal analysis, such as the organic analysis in Chemistry, and in the exact sciences by the application of the principles of demonstrative proof.

Passing by, only because our limits do not permit their consideration, the system of Bacon, which was adopted by the French encyclopædists of the last century, and the details of the system of Locke, we come down to our own times before we find any definite attempt to supply the want. An eminent Scotch writer, as he reviewed the efforts of previous philosophers to classify human knowledge, asserted that it was an impossible task, and so, from its magnitude, it would fairly seem.

Nothing daunted by such an assertion, Coleridge suggested the plan of classification which was adopted in the arrangement of the English "Encyclopædia Metropolitana," but which he found to require, after he had exhausted his categories, an additional category of "Miscellaneous" species— the unfortunate subalterns which had no summum genus under which to range themselves.

Among the curious but highly philosophic remains of Jeremy Bentham is a proposed system of scientific classification; but, like his other works, it is only a storehouse of theory from which less gifted but more practical men draw capital for constant use.

All the more modern writers agree in considering the system of *Ampère* the most correct and useful. It is based upon the two categories of *mind* and *matter*, and under these it expands into a very great number of subordinate sciences, many of which, it must be said, are created, *i. e.*, in name, to fill up gaps which would spoil the symmetry of his table.

It is not our purpose to write out his table in full; it would be out of place in a text-book, as it could only be examined, not studied; but we will form a tree of one or two of his subjects to illustrate his plan and indicate its truthfulness and use.

His *First Table* contains:

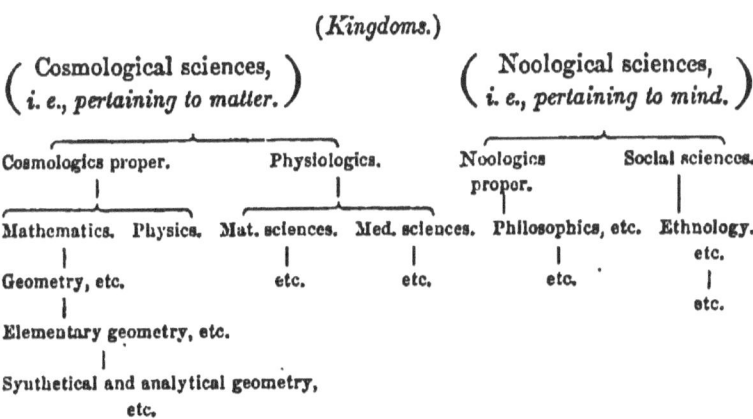

Of these there are several tables and more than a hundred branches. In thus indicating rather than writing out in full the tables of Ampère, we spare the student the reading, in place, of many names unknown to our ordinary scientific studies, such as *Dialegmatics, Eleutherotechnics, Technesthetics,* while we present to him what is alone our present purpose, the theory and principle of classification.

The chief merit of his tables, which he spent his life in constructing, seems to be that there are no cross divisions—that no subordinate science lies out of its own class or laps over into another—errors which rendered Bacon's system worthless, and which caused Bentham to abandon his great idea and leave it in its inchoate form.

Auguste Comte, who has given to the world, in his *Cours de la Philosophie Positive*, his views of philosophy, did not attempt so much to classify science as to determine the true relation between general science and positive science—to make positive science more general in its application and general science more practical and positive. This has been his life-work. There is much of his work which bears indirectly but dangerously upon religious belief, and there is an elaborate description of the historical progress of positive science through what he calls the *mystical* and *metaphysical* eras to the *positive*.

To explain more clearly his view of this *positive* era, it is that in which the *mysticism* or *mythology* of ancient and early times, as well as the crude *metaphysical* notions of the Middle Ages which found their issue in astrology and magic, are swept away by the light of modern free thought and investigation, and in their place are substituted the laws of creation—laws which regulate its origin, its progress and its destiny. There are six *positive* sciences which include everything that can be known. These are *Mathematics, Astronomy, Physics, Chemistry, Biology* and *Sociology*.

But it is not within our scope to explain his philosophy; we have only to do with its Logic, and this is found in his classification.

The subject of classification is yet open, and will become, without doubt, clearer and more practical as science advances to the discovery of the proximate laws of creation.

(61.) Conclusion.

From the foregoing investigation of the art of Reasoning, we may pause a moment at the end to reflect upon its real value and importance. If Logic is really the art which controls and guides the reason in its workings, and without which we can attain to no truth upon which the reason is exercised, it is surely worthy of a high place in the catalogue of elementary studies, and the statement and adoption of its laws must be considered of the first importance.

And, above all, should it be placed upon its own foundation, and dissociated from any other sciences which either rob it of its own identity or use it without acknowledging its office.

APPENDIX.

EXAMPLES FOR PRAXIS.

LOGICAL *praxis* consists in the application of the rules of Logic as a test of all the forms of argument. The following examples for praxis are designed to give ease and logical quickness of detection to the student. They comprise illustrations of all kinds and forms of argument—regular syllogisms, irregular and inverted arguments, compound arguments, fallacies of every kind, curious propositions, examples of the processes of generalization and division, amphibolous sentences, etc., etc. A certain number of these should be given to the student, as an exercise with each lesson, upon the review of the subject. He should be required to state what each is in its present form—if a *fallacy*, of what kind; if a *logical fallacy*, to write it out by symbols and thus to expose its invalidity; if an *inverted argument*, to put it in the true order of sequence of premiss and conclusion; if an *enthymeme*, to supply the suppressed premiss; if in an *imperfect* mood, to reduce it to one of the *perfect moods of the first figure*,—in a word, to show by this practice the truth of the assertion made at the beginning of this book, and steadily kept in view throughout the work, that every valid argument, whatever its form, may be brought directly to the dictum of Aristotle as the final test of argument.

In a few of the more difficult examples, to guide the student, a reference has been made to the page on which their type may be found. Some selected arguments from the Latin authors, generally read in the schools, have been added, as of interest to the classical student.

1. Jupiter, Saturn, Venus, Earth, etc. move round the sun in ellipses; these are all planets; therefore all planets move round the sun in ellipses.

2. Induction is the only true science of reasoning; Syllogistic Logic is not induction; therefore Syllogistic Logic is not a true science of reasoning.

3. No one is good who commits sin; all men commit sin; therefore there is none good except God.

4. A story is not to be believed the reporters of which give contradictory accounts of it; the story of Napoleon's life is of this kind; therefore it is not to be believed.

5. Every one desires happiness; virtue is happiness; therefore every one desires virtue.

6. No evil should be allowed that good may result; all punishment is an evil; therefore no punishment should be allowed.

7. Those who are over-credulous should not be believed; the ancient historians were over-credulous; therefore we should believe nothing they say.

8. An American citizen should be free; I am an American citizen; therefore I should be allowed to do whatever I please.

9. The duke yet lives that Henry shall depose. (v. p. 154.)

10. All the peaches in this field are worth one hundred dollars; this is one of the peaches in this field; therefore it is worth one hundred dollars.

11. Ought we to act from expediency as a motive?

12. Ought not children to obey their parents?

13. A designing character is not worthy of trust; therefore I do not trust engravers.

14. All good men are beloved by their associates; this man is beloved by his; therefore he must be good.

15. ———— Pallas ne exurere classem
 Argivum atque ipsos, potuit submergere ponti?
* * * * * * *

> Ast ego quæ Divum incedo regina Jovisque
> Et soror et conjux, una cum gente tot annos
> Bella gero.

16. Happiness consists in obedience to the Divine Laws; this obedience is virtuous conduct; virtuous conduct is the subordination of the inferior to the superior in our nature; this subordination is induced by self-control; therefore happiness is the result of self-control.

17. Crime is a violation of the laws of our country; piracy is crime; this man belongs to a band of lawless men, and this band has been taken in the very deed of piracy; therefore he has violated the laws of his country.

18. He that is of God heareth my words; ye therefore hear them not, because ye are not of God.

19. We must do one of three things—go back, stand still, or go forward; we cannot go back or stand still; therefore we must go forward.

20. "Ay, in the catalogue ye go for men—
> As hounds and greyhounds, mongrels, spaniels, curs,
> Shoughs, water-rugs and demi-wolves are called
> All by the name of Dogs."

21. All that glitters is not gold; tinsel glitters; therefore it is not gold.

22. Warm countries alone produce wine; therefore Spain produces wine.

23. Quo melior servo quò liberior sit avarus,
> In triviis fixum, cum se demittit ob assem,
> Non video. Nam qui cupiet, metuet quoque porro
> Qui metuens vivit, liber mihi non erit unquam.

Or, The fearful man is not free; the miser is fearful; therefore the miser is not free.—Hor. Ep. 1, 16.

The following strong eulogium of Logic is an argument of the schoolmen, who called it " The Divine art; the eye of the Intellect; the art of arts; the science of sciences; the bulwark of philosophy":

24. Utque supra Æthereos sol aureus emicat ignes,
Sic artes inter prominet hæc Logica;
Quid? Logica superat solem; sol namque, diurno
Tempore dat lucem, nocte sed hancce negat;
At Logicæ sidus nunquam occidit; istud in ipsis
Tam tenebris splendet, quam redeunte die.

Cum hoc, ergo propter hoc, a form of the *non causa pro causa,* is broadly illustrated by the following:

25. The encroachment of the sea upon that bank upon the coast of Kent known as the Goodwin Sands, rendering it very dangerous to navigation, led to the appointment of a committee of parliament to inquire into the subject. The committee went down, and examined, among other witnesses, an old man, who, when asked what he regarded as the cause of this encroachment, replied, after some minutes' thought, that he did not know, unless it had something to do with Tenterden steeple, as he remembered nothing of the kind before they began to build that steeple, but it had been steadily growing worse ever since.

26. Horses are stronger than men; elephants are stronger than horses; therefore elephants are stronger than men.

27. Men need the restraints of government, because they have vicious propensities.

28. Unjust laws endanger the stability of government, because (———); laws which enslave man's conscience are unjust because (———); therefore laws which restrain the freedom of conscience endanger the stability of government.

29. If we suppose the telegraphic connection from London to be made around the world, and the transmission to be instantaneous, then a message starting from London at 12 o'clock to-day would reach London at 12 o'clock yesterday.

30. If men are to be punished hereafter, God must be the punisher; if God be the punisher, the punishment must be just; if the punishment is just, the punished must be guilty; if they are guilty, they could have acted otherwise; if they

could have acted otherwise, they were free agents; therefore, if men are liable to punishment in another world, they must be free agents.

31. This medicine cured a very difficult case of disease, therefore it will cure every disease.

32. Among the most bitter persecutions known to history were those of the French Revolution; therefore they must have been religious persecutions.

33. Testimony is likely to be false; the existence of the Pyramids depends on testimony; therefore we may doubt whether there are pyramids in Egypt.

34. No man can perform impossibilities; a miracle is an impossibility; therefore no man can perform a miracle.

35. With God all things are possible.

36. No man can do these miracles which thou doest, except God be with him.

37. Si testibus credendum sit contra argumenta, sufficit, tantum judicem esse non surdum.—*Bacon's Antitheta.*

38. Hæc, si displicui, fuerint solatia nobis;
    Hæc fuerint nobis præmia, si placui.—*Martial.*

39. From the existence of bad morals springs the making of good laws; from good laws arises the safety of the commonwealth; from the safety of the commonwealth all social good things flow; therefore from the existence of bad morals come all good things to society.

40. Si saperem odissem jure sorores,
    Numina cultori perniciosa suo,
    At nunc (tanta meo comes est insania morbo),
    Saxa memor refero rursas ad icta pedem.—*Ovid.*

41. Cæsar oppressit patriam; Tullius non oppressit patriam; ergo (———).

42. Una Eurusque; notusque ruunt, creberque procellis, Africus.

43. For whom he did foreknow, he did also predestinate

to be conformed to the image of his Son; that he might be the first born among many brethren. Moreover, whom he did predestinate them he also called; and whom he called them he also justified; and whom he justified them he also glorified. Rom. viii. 29, 30.

44. When the sun is in Cancer, it is summer; it is now summer; therefore (———).

45. All persecution for conscience' sake is unpleasing to God, because it is injustice.

46. Genius must join with study to make a great man; this man will never be great, for, though he has genius, he cannot study.

47. No man can serve two masters.——Ye cannot serve God and mammon.

48. Pride and innocence are incompatible. The angels are innocent; therefore (———).

49. In this life we must either obey our vicious inclinations or resist them; if we obey them, we shall have sin and sorrow; if we resist them, we shall have pain and labor; therefore we cannot be free from trouble in this life.

50. This doctrine cannot be proved from the Gospels; nor from the Acts of the Apostles; nor from Epistles; nor from the Revelation of St. John; therefore it cannot be proved from the New Testament. (v. p. 175.)

51. It is a sin to kill a man; a murderer is a man; therefore he should not be hanged.

These examples may be increased at the pleasure of the teacher. The author would suggest that it would be well for students, in their readings both of verse and prose, and in their classical studies as well as in English, to cultivate a habit of marking the different logical forms of discourse. It would soon become a pleasant pastime, as well as a profitable lesson.

www.ingramcontent.com/pod-product-compliance
Lightning Source LLC
Chambersburg PA
CBHW031812230426
43669CB00009B/1116